Lecture Notes in Mathematics 1922

Avner Friedman (Ed.)

Tutorials in Mathematical Biosciences IV

Evolution and Ecology

With Contributions by:

C. Cosner · D. Janies · L.S. Kubatko
Y. Lou · T. Nagylaki

 Springer

Editor

Avner Friedman
Mathematical Biosciences Institute
Ohio State University
231 West 18th Avenue
Columbus, OH 43210-1292
USA

e-mail: afriedman@math.ohio-state.edu
afriedman@mbi.ohio-state.edu

Library of Congress Control Number: 2007933684

Mathematics Subject Classification (2000): 35J20, 35J60, 35K55, 35K57, 62P10, 62P12, 92B99, 92D10, 92D15, 92D25, 92D40

ISSN print edition: 0075-8434
ISSN electronic edition: 1617-9692
ISBN 978-3-540-74328-6 Springer Berlin Heidelberg New York
DOI 10.1007/978-3-540-74331-6

Springer is a part of Springer Science+Business Media
springer.com
© Springer-Verlag Berlin Heidelberg 2008

Typesetting by the author and SPi using a Springer LaTeX macro package

Cover design: *design & production* GmbH, Heidelberg

Printed on acid-free paper   SPIN: 12109630   41/SPi   5 4 3 2 1 0

Preface

This is the fourth volume in the series "Tutorials in Mathematical Biosciences." These lectures are based on material which was presented in tutorials or developed by visitors and postdoctoral fellows of the Mathematical Biosciences Institute (MBI), at The Ohio State University. The aim of this series is to introduce graduate students and researchers with just a little background in either mathematics or biology to mathematical modeling of biological processes. The first volume was devoted to mathematical neuroscience, which was the focus of the MBI program 2002–2003. The second volume dealt with mathematical modeling of calcium dynamics in signal transduction, the focus of the MBI program in the winter of 2004. The third volume dealt with topics of cell cycle, tumor growth, and cancer therapy; these topics featured in several workshops held at the MBI in the fall of 2003. The present volume deals with a variety of topics of evolution and ecology, which were considered in the MBI during the year 2005–2006. These topics include phylogenetics; evolution of genes through migration–selection; ecological modeling; and evolution of dispersal and population dynamics. Documentation of the 2005–2006 activities, including streaming videos of the workshops, can be found on the Web site: http://mbi.osu.edu.

Phylogenetics is the study of the evolutionary relations of genes and organisms. Phylogenetic trees are represented by graphs in which the leaves represent observed biological entities. In constructing such graphs, one tries to trace the evolution of species, traits, or diseases. The first two chapters of this volume deal with phylogenetics. Chapter 1 is a general survey on estimation of phylogenetic trees with emphasis on likelihood methods. Chapter 2 is concerned with computational methods of very large trees, exploring other optimality methods, with application to the study of the evolution of SARS and influenza.

The next three chapters deal with population genetics and population dynamics. Chapter 3 introduces reaction–diffusion equations as a mathematical framework to study ecological models. It then addresses the following ecological questions: what is the minimal patch size necessary to support a population?; when do biological invasions occur?; and what spatial patterns can form?

Chapter 4 focuses on evolution and genes. The genetic composition of a population is described by genotypic or allelic frequencies, using either deterministic models or stochastic models. The models presented here are both discrete and continuous. The questions discussed include the loss, or the maintenance, of a specified allele, and the stability of completely polymorphic equilibria.

The final chapter is concerned with the effects of dispersal and spatial heterogeneity on population dynamics, via reaction–advection–diffusion models. Issues regarding how advection along resource gradients affect the extinction of species or how invasion of rare species may take place are considered.

It is not uncommon to see the same biological processes benefit by using different mathematical and statistical approaches. This volume is a good example: Although the mathematical and statistical tools developed or reviewed here are quite varied, the biological themes have a common thread as they all deal with the evolution of species in an evolving ecological system.

I express my appreciation and thanks to Daniel Janies, Diego Pol, Laura Salter-Kubatko, Thomas Nagylaki, Yuan Lou, and Chris Cosner for their marvelous contributions. I hope this volume will serve as a useful introduction to those who want to learn about important and exciting problems that arise in evolution and ecology.

Contents

1

Inference of Phylogenetic Trees

L.S. Kubatko

Departments of Statistics and Evolution, Ecology, and Organismal Biology,
The Ohio State University, Columbus, OH 43210, USA
email: lkubatko@stat.osu.edu

Study of the evolutionary relationships among organisms has been of interest to scientists for over 100 years. The earliest attempts at inferring evolutionary relatedness relied solely on observable species characteristics. Modern molecular techniques, however, have made available an abundance of DNA sequence data, which can be used to study these relationships. Today, it is common to consider the information contained in both types of data in order to obtain robust estimates of evolutionary histories.

These evolutionary histories are most commonly represented by a phylogenetic tree, which is mathematically described as an acyclic connected graph (V, E), where V is the set of vertices and E is the set of edges. Vertices connected through only a single edge are called terminal nodes, while vertices connected by more than one edge are called internal nodes. In phylogenetic tree reconstruction, it is common to assume that trees are bifurcating, so that each internal node is connected through exactly three edges, with the exception that for a rooted tree the root is connected through two edges.

Estimation of the phylogenetic relationships among a collection of organisms given genetic data for these organisms can be divided into two distinct problems. The first is to define the particular criterion by which we compare the fit of a particular phylogenetic hypothesis to the observed data. The second is to search the space of possible phylogenies for the particular tree or trees that provide the best fit to the data. In this chapter, we give an overview of these two problems, with particular emphasis on the maximum parsimony and maximum likelihood criteria for comparing trees. Techniques for searching the space of trees for optimal phylogenies under these criteria are also discussed. Throughout the chapter, we use two data sets to illustrate the main ideas. We begin by defining some of the commonly used terminology, and by providing a careful description of the data used in phylogenetic analysis.

1.1 Introduction and Terminology

1.1.1 Phylogenetic Trees

As described earlier, a phylogenetic tree can be viewed as a graph for which the terminal nodes represent organisms for which data are observed, called *taxonomic units* or *taxa*, while the internal nodes represent hypothetical ancestral organisms. The edges connecting the nodes are generally referred to as branches and denote ancestry-descent relationships. Often, the lengths of the branches are taken to represent evolutionary time. In this chapter, the word *topology* will be used to refer to the labeled branching pattern of a tree without regard to branch lengths.

Phylogenetic trees are called rooted when the location of the common ancestor of all the taxa in the tree is identified, or unrooted when no such common ancestor is specified. Rooted phylogenetic trees may or may not satisfy the assumption of a molecular clock. The molecular clock hypothesis is that the rate of evolution is approximately constant over time. When all of the sequences in the tree are contemporaneous, this assumption restricts the lengths of the branches so that the sum of the branch lengths connecting each taxon to the root is the same for all taxa. Examples of phylogenetic trees are shown in Fig. 1.1.

As the number of taxa under consideration grows, the number of distinct topologies increases rapidly. For n taxa, the number of unrooted labeled bifurcating topologies is

$$\prod_{i=3}^{n} (2i - 5).\tag{1.1}$$

An unrooted topology has $n - 2$ internal nodes and $2n - 3$ branches. Because adding a root to an n-taxon tree amounts to placing it along any of the $2n - 3$ branches, the number of rooted topologies for n taxa is the found by applying (1.1) for $n + 1$. A rooted topology for n taxa contains $n - 1$ internal nodes and $2n - 2$ branches. Table 1.1 shows the rapid increase in the number of topologies as a function of the number of taxa.

Table 1.1. Number of topologies, internal nodes, and branch lengths for unrooted bifurcating topologies as a function of the number of tips in the tree

Number of tips	Number of topologies	Number of internal nodes	Number of branches
5	15	3	7
10	2,027,025	8	17
20	2.2164×10^{20}	18	37
50	2.8381×10^{74}	48	97

Table 1.2. DNA sequences for a portion of the *L1* gene for seven Group A9 human papillomaviruses

HPV16	ATGTGGCTGCCTAGTGAGGCCACTGTCTACTTGCCTCCTGTCCAGTATCTAAGGTTG
HPV35h	ATGTGGCGGTCTAACGAAGCCACTGTCTACCTGCCTCCAGTTCAGTGTCTAAGGTTG
HPV31	ATGTGGCGGCCTAGCGAGGCTACTGTCTACTTACCACCTGTCCAGTGTCTAAAGTTG
HPV52	ATGTGGCGGCCTAGTGAGGCCACTGTGTACCTGCCTCCTGTCCTGTCTCTAAGGTTG
HPV33	ATGTGGCGGCCTAGTGAGGCCACAGTGTACCTGCCTCCTGTCCTGTATCTAAAGTTG
HPV58	ATGTGGCGGCCTAGTGAGGCCACTGTGTACCTGCCTCCTGTCCTGTGTCTAAGGTTG
RhPV1	ATGTGGCGGCCTAGTGACTCCAAGGTCTACCTACCACCTGTCCTGTGTCTAAGGTGG

See Sect. 1.1.3 for a description of the data.

1.1.2 Data for Phylogenetic Estimation

The most common type of data used in phylogenetic inference is discrete character data. These data can be represented by a matrix X in which entry x_{ij} represents the particular state of the character observed for taxon i at position j. Both morphological data and molecular sequence data (DNA, RNA, amino acid, or protein sequences) are examples of discrete character data. For data of this type, we assume that each character (column) in the data matrix is homologous, which means that in each of the taxa the particular state observed was derived from an ancestral state that was common to all of the taxa in the data matrix. In practice, assessment of homology can be difficult, particularly for molecular sequence data.

Development of a data matrix for use in phylogenetic inference for molecular sequence data is a nontrivial task, because the molecular sequences are derived individually for each taxon and must subsequently be placed into a data matrix so that the assumption of homology is likely to be satisfied. The process of constructing the data matrix for a collection of taxa is called *sequence alignment*. Table 1.2 shows an example of an aligned portion of the *L1* gene for seven human papillomaviruses. In this chapter, the problem of sequence alignment will not be discussed, and we will assume that the data have already been aligned. The interested reader is referred to several references on the topic: [54, 64, 71, 82, 83].

1.1.3 Example Data Sets

Throughout this chapter, we will use two data sets to illustrate several techniques for phylogenetic inference. The first is a set of viral sequences for a particular gene, and the second consists of both morphological and molecular data on cephalopods. Further details concerning each data set are given below.

Papillomaviruses

Papillomaviruses are a group of viruses that infect a variety of organisms ranging from birds to mammals, including humans. They are small nonenveloped

DNA viruses that generally cause benign epithelial lesions, though some types may cause malignancies. The papillomavirus genome is approximately 8,000 base pairs in length and is divided into an early region (E), which encodes genes expressed immediately after infection of the host, and a late region (L), which encodes two capsid proteins. The early region comprises over 50% of the genome, and contains six open reading frames (E1, E2, E4, E5, E6, and E7). The late region comprises approximately 40% of the genome and encodes two proteins, L1 and L2. The remaining 10% of the genome is a long control region (LCR) that does not code for proteins but does contain transcription factor binding sites and the origin of replication.

Papillomaviruses are classified into types, subtypes, and variants based on the sequence of the *L1* gene. They are also grouped based on sequence similarity, host type, and pathogenic characteristics. In this example, we consider the sequence of the *L1* gene for thirty Group A papillomaviruses, 28 of which infect humans. Notable among this collection of sequences are human papillomavirus (HPV) types 16, 18, and 31, which are found to be associated with over 95% of cervical cancer cases [87]. The particular sequences studied here, as well as their genetic subtype and pathology, are listed in Table 1.3. More information on these particular sequences can be found in Ong et al. (1997), and information concerning the genetics of papillomaviruses in general can be found in Zheng and Baker (2006).

For this example, aligned DNA sequences were downloaded from the HPV Database maintained by Los Alamos National Labs (http://hpv-web.lanl.gov/). This alignment was edited by limiting the analysis to only the 30 taxa in Table 1.3, removing the sequence prior to the start codon in all taxa, and removing all sites for which all of the taxa had an insertion or deletion, resulting in 1,560 aligned sites.

Cephalopods

Cephalopods (e.g., squids, cuttlefishes, octopi) are a diverse class of molluscs containing over 800 species. They inhabit a wide range of marine environments, from coastal to benthic waters, and vary in size from 10 mm to several meters. Taxonomically, the class is divided into two groups, Nautiloidea and Coleoidea. Nautiloidea contains only a single genera, while Coleoidea contains all remaining extant taxa. Three subgroups within Coleoidea are recognized: Decabrachia (squids and cuttlefishes), Octobrachia (octopi), and Vampyromorpha. The placement of Vampyromorpha has been controversial, with some analyses supporting a sister relationship with Octobrachia and others placing Vampyromorpha with Decabrachia [6, 8].

For this example, we consider a subset of the data examined by Lindgren et al. [46], which includes both molecular and morphological data for 78 molluscs. Fifteen taxa, including representative taxa from the Decabrachia and

Table 1.3. Group A papillomaviruses, genetic subtypes, risk classification [38], and host tissue infected for the virus types studied here

Genetic subtype	Group	Risk classification	Host tissue infected
HPV32	A1	Low	Oral
HPV42	A1	Low	Genital
HPV3	A2	Low	Cutaneous
HPV10	A2	Low	Cutaneous
HPV2a	A4	Low	Cutaneous, mucousal
HPV27	A4	Low	Cutaneous, genital
HPV57	A4	Ambiguous	Oral, genital
HPV26	A5	Ambiguous	Cutaneous, possibly genital
HPV51	A5	High	Genital
HPV30	A6	Low	Cutaneous, mucousal
HPV53	A6	Ambiguous	Genital
HPV56	A6	High	Genital
HPV18	A7	High	Genital
HPV45	A7	High	Genital
HPV39	A7	High	Genital
HPV59	A7	High	Genital
HPV7	A8	Low	Cutaneous, oral
HPV40	A8	Low	Genital
HPV16	A9	High	Genital
HPV35h	A9	High	Genital
HPV31	A9	High	Genital
HPV52	A9	High	Genital
HPV33	A9	High	Genital
HPV58	A9	High	Genital
RhPV1	A9	Unclassified	Genital
HPV6b	A10	Low	Oral, genital
HPV11	A10	Low	Oral, genital
HPV13	A10	Low	Oral, genital
PCPV1	A10	Unclassified	Oral
HPV34	A11	Low	Oral, genital

Octobrachia as well as Vampyromorpha, are considered here, for both the morphological data assembled by Lindgren et al. (2004) and for three nuclear genes, 18S (3,477 sites, of which 808 are parsimony informative), 28S rRNA (667 sites, of which 238 are parsimony informative), and histone H3 (327 sites, of which 73 are parsimony informative), that they examined. The taxa selected for analysis are shown in Table 1.4. This data set will be used to highlight the differences in analyzing molecular and morphological data in a phylogenetic context.

Table 1.4. Cephalopod taxa included in our examples (from Lindgren et al. (2004))

Group	Species name	Type of cephalopod
Octobrachia		
	Stauroteuthis syrtensis	Cirrate octopus
	Thaumeledone guntheri	Benthic octopus
Vampyromorpha		
	Vampyroteuthis infernalis	Vampire quid
Decabrachia		
	Sepia officinalis	Cuttlefish
	Heteroteuthis hawaiiensis	Bobtail squid
	Spirula spirula	Ram's horn squid
	Idiospeius pygmaeus	Pygmy squid
	Loligo pealei	Common market squid
	Architeuthis dux	Giant squid
	Enoploteuthis leptura	Open ocean squid
	Pyroteuthis margaretifera	Open ocean squid
	Gonatus fabricii	Open ocean squid
	Histioteuthis hoylei	Open ocean squid
	Ommastrephes bartrami	Open ocean squid
	Psychroteuthis sp.	Open ocean squid

1.2 Optimality Criteria

Given a data matrix X consisting of either aligned molecular sequences or morphological data, it is necessary to develop methods for constructing a phylogenetic tree that appropriately represents the information concerning evolutionary relationships contained in X. There are three general classes of methods for constructing phylogenies from a given data matrix. The first set of methods are distance methods, in which the original data matrix X is first converted to a matrix of pairwise distances between taxa, and these distances are used to construct the phylogeny. Distance methods will not be considered further here, but see [45, 55, 68, 77] for details.

The second two methods, parsimony and maximum likelihood, are based on the definition of a criterion for comparing alternative trees. The problem of constructing a phylogeny from a data matrix is then reduced to two smaller problems. The first is the evaluation of the selected optimality criterion for any particular tree, and the second is the search over the large space of trees for the particular tree that optimizes the selected criterion. In this section, the parsimony and likelihood criteria are discussed, and methods for computing the scores of individual trees are described. The problem of searching for optimal trees will be considered in the next section.

1.2.1 Parsimony

Parsimony, one of the most common methods for inferring phylogenies, is also one of the oldest, dating back to its introduction by Edwards and Cavalli-Sforza [11] in 1964 (see Chap. 10 in Felsenstein [20] for a nice account of the history of the field of phylogenetics). The parsimony method in phylogenetics is based on the general principle that simpler hypotheses should be preferred over more complex ones, where "simplicity" in the phylogenetic context is translated to mean the least amount of evolutionary change. Thus, trees that minimize the total amount of evolutionary change for a given data set are preferred, and the tree requiring the minimum number of evolutionary changes to explain the given data is called the most parsimonious or maximum parsimony (MP) tree.

Because the parsimony criterion is concerned with minimizing the amount of postulated evolutionary change, it can be applied to a variety of genetic data – essentially all that is required is a mechanism for quantifying "evolutionary change" in the observed data. The criterion can then be evaluated for any given tree by computing the amount of change required by that tree for the observed characters. To be more precise, consider a particular character, say x, and let x_i^h be the state of character h at node i in the tree, $1 \leq i \leq 2n - 2$, where nodes 1 through n are external nodes corresponding to the tips of a rooted tree for which the character states are observed, and nodes $n+1$ through $2n-2$ are internal nodes whose character states must be inferred. Define $C(x_i^h, x_j^h)$ to be the cost of changing from the state for character h at node i to the state for character h at node j over the branch connecting nodes i and j. Note that it does not have to be the case that $C(x_i^h, x_j^h) = C(x_j^h, x_i^h)$, though equality is commonly assumed. The parsimony score of a tree, τ, under this criterion is then given by

$$S(\tau) = \sum_{h=1}^{N} \sum_{b=1}^{B} C(x_{b_1}^h, x_{b_2}^h), \qquad (1.2)$$

where N is the number of characters in the data set, B is the number of branches in the tree, and b_1 and b_2 are the nodes at the ends of branch b, for which either the character state has been observed, or an optimal character state has been assigned. From (1.2), we see that the length of a tree is computed by summing lengths over all branches for a particular character in the data matrix, and then summing over all characters in the data matrix. Performing the calculation in this way necessarily assumes that changes among states on the branches occur independently once the states at all nodes are known, and also that changes across characters are independent. A weight, w_h, can be added in front of the cost term in (1.2) to allow for differential weighting of the characters in a data matrix.

The most commonly used cost function is one in which $C(x_i^h, x_j^h)$ is 1 if $x_i^h = x_j^h$ and 0 otherwise. This cost function counts the number of changes between character states in the tree and weights each type of change equally, i.e., any differences in state at the two ends of a branch increase the score of the tree by one, regardless of what those states are. This cost function can be applied to unordered multistate data, which can include molecular data (nucleotide and amino acid) as well as morphological data. This is generally referred to as Fitch parsimony. However, many variations of this cost function are possible. For example, for nucleotide data, it may be sensible to assign a lower cost to transitions than transversions, while for amino acid data, we might assign different costs for synonymous vs. nonsynonymous changes. For morphological data, we might specify an ordering in the data. For example, it is unlikely for beak size to change from "small" to "large" without first being "medium," and so an observed change from "small" to "large" might incur a cost equal to the sum of the costs of changing from "small" to "medium" and "medium" to "large." This is an example of what is known as Wagner parsimony [41, 18], for which costs are assigned to ordered multistate data in such a way that a change from one state to another incurs the sum of the costs of any intervening states.

To illustrate the computation of the parsimony length of a tree, consider the morphological data for the cephalopod example. We consider two trees representing alternative hypotheses concerning the placement of Vampyromorpha as described in Lindgren et al. (2004). These trees are shown in Fig. 1.1, and the observed states for character 38 in the data matrix are given at the tips of the trees (the complete data set is given in Lindgren et al. (2004)). Note that for this character, all species of Decabrachia have state 1, and thus the length of the clade containing the Decabrachia is 0. Thus, the Decabrachia clade has been collapsed into a single node in the trees in Fig. 1.1. To calculate the length of the trees, consider first the tree in Fig. 1.1a, and consider the node ancestral to *S. syrtensis* and *T. guntheri*. Since both *S. syrtensis* and *T. guntheri* have state 0, this ancestral node can also be assigned state 0, and this assignment requires no changes along the branches descending from it. Next, consider the node ancestral to *V. infernalis* and the ancestor of *S. syrtensis* and *T. guntheri*. Looking at this node's two descendants, we see that one of them (*V. infernalis*) has state 1 and the other (the ancestor of *S. syrtensis* and *T. guntheri*) has state 0. In this case, we could assign either state as a possible ancestral state. Note that selection of either state as the ancestral state will lead to one change along the tree, and thus we increase the length of the tree by one. Moving to the next most ancestral node, we see that the Decabrachia have state 1 and thus an assignment of state 1 to the ancestor of the Decabrachia and the clade containing *V. infernalis*, *S. syrtensis*, and *T. guntheri* is most parsimonious, and does not increase the length of the tree. Finally, the node at the root of the tree can be assigned state 1 without increasing the length of the tree. The overall length of this tree for this character is then 1.

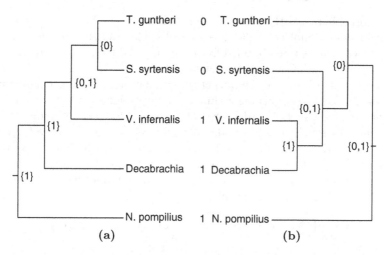

Fig. 1.1. Trees representing distinct hypothesized relationships among cephalopods, taken from Lindgren et al. (2004). The tree in (**a**) is the consensus of the nine MP topologies for the morphological data (each MP tree has length 107; the consensus tree has length 109). The tree in (**b**) is supported by the molecular data. Sets at the internal nodes of the tree are used to compute the score of the tree under the Fitch algorithm (see text for details)

This algorithm for computing the length of a tree, called the Fitch algorithm [26], can be expressed in more mathematical terms as follows. For each node in the tree, a set of character states will be assigned. The set at the tips of the tree contains a single state, the observed state at that tip. Then, for any node for which a character state set has been assigned for its two immediate descendants, the state set assigned to that node is the intersection of the state sets of its two immediate descendants if that intersection is nonempty; otherwise, it is the union of the state sets of the two immediate descendants. Whenever a union of state sets is required, the length of the tree is increased by one. The Fitch algorithm was developed specifically for unordered multistate characters, such as nucleotide and protein data, for which any state can change directly to any other state. Since changes in either direction are weighted equally under this method, a tree can be arbitrarily rooted with no change to its length, which allows one to root the tree at the most convenient location.

Comparing the trees in Fig. 1.1a,b, we see that for this character, the tree in Fig. 1.1a has length 1 and the tree in Fig. 1.1b has length 2, and so the tree in Fig. 1.1a is preferred for this character. Of the 45 parsimony informative characters in this data set, nine are informative for selecting between these two trees. Of these nine, eight favor tree (a) (characters 10, 38, 40, 45, 49, 57, 59, and 60) and one (character 6) favors tree (b). The result is that analysis of the morphological data favors placement of Vampyromorpha with the Octobrachia rather than the Decabrachia, which conflicts with the results obtained from the molecular data, as we will see in later sections. Figure 1.1a shows the consensus tree obtained from the nine MP trees (each MP tree has length 107; the consensus tree has length 109).

A second algorithm for computing the score of a tree under parsimony is the Sankoff algorithm. This algorithm works by assigning a function to each node of the tree which records, for each possible state, the minimum score for the subtree rooted by that node. We denote this function by $S_i^h(x)$, and define it to be the minimum score for the subtree rooted by node i assuming that node i has state x for character h. This value can be computed for any node for which this function has already been computed for its two immediate descendants using the following relationship

$$S_i^h(x) = \min_{x_j^h}\{C(x_i^h, x_j^h) + S_j^h(x_j^h)\} + \min_{x_k^h}\{C(x_i^h, x_k^h) + S_k^h(x_k^h)\}, \qquad (1.3)$$

where j and k are the two nodes directly descending from node i. This equation is very intuitive. For example, consider the first term. This term corresponds to the branch descending from node i to node j. This branch contributes to the length of the subtree descending from node i in two ways: first, it contributes a length along the branch connecting nodes i and j; second, it contributes a length due to the subtree descending from j, as recorded by the S function for node j. There is then a similar contribution from the other branch descending from node i, denoted by k here. Taking the minimum over all possible assignments of states to the nodes j and k will give the minimum at node i, given that it has state x.

This algorithm is applied successively to the nodes of the tree in a postorder traversal (see Felsenstein ([20], p. 587)). The value of the S function at the tips of the tree is determined by setting $S_m^h(x) = 0$ if tip m has state x for character h, and $S_m^h(x) = \infty$ otherwise. The minimum length of the entire tree is then found by selecting the minimum value of the S function at the root of the tree. Denoting the root node by r, the parsimony score of the tree is

$$S(\tau) = \sum_{h=1}^{N} \min_x S_r^h(x). \qquad (1.4)$$

Figure 1.2 gives an example of the computation for the morphological data for the cephalopod example, with both the simple cost matrix used in the explanation of the Fitch algorithm, and a modified cost matrix that results in a different conclusion concerning which tree (of the two) is the most parsimonious.

We note that both the Fitch and Sankoff algorithms are dynamic programming algorithms, since they reduce the problem of computing the score to subproblems, which can be optimally solved in such a way that it can be proved that they lead to the overall optimal solution. Felsenstein ([20]; p. 16) discusses the connection between the two methods. The Sankoff algorithm is

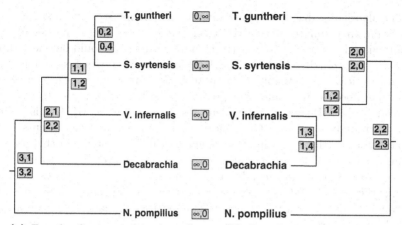

(a) For the first cost function, $S = 1$, while for the second cost function, $S = 2$.

(b) For both cost functions, $S = 2$.

Fig. 1.2. Trees representing distinct hypothesized relationships among cephalopods, taken from Lindgren et al. (2004). The tree in (**a**) is the consensus of the nine MP topologies for the morphological data. The tree in (**b**) is supported by the molecular data. The colored boxes at the nodes of the tree represent the $S()$ function used to compute the length of tree under the Sankoff algorithm for two different cost functions. The upper (*blue*) boxes at each node correspond to the same cost function as was used to illustrate the Fitch algorithm: $C(0,0) = C(1,1) = 0$; and $C(0,1) = C(1,0) = 1$. The lower boxes (*yellow*) correspond to a cost function that penalizes more for one particular change: $C(0,0) = C(1,1) = 0; C(0,1) = 1$; and $C(1,0) = 2$. For the first cost function, the tree in (**a**) is preferred, while for the second cost function, the scores of the two trees are equivalent

more general, in that it allows the use of *any* cost function, while the Fitch algorithm is confined to the setting where all changes are weighted equally. We also note that while both algorithms specify a sum over characters to compute the total score for the tree, the computation can be simplified for both algorithms by computing the scores for only unique sites. For example, any character for which all taxa have the same state will require no changes on every tree. Additionally, under Fitch parsimony, any character for which all taxa except one have the same state will require exactly one change on any tree. Characters of this nature are generally said to not be *phylogenetically informative*, since they do not prefer any tree over any other in the parsimony setting. Therefore, no computations need be performed on these character patterns. However, these character patterns do contribute to estimation in other settings, as will be seen for likelihood in the following section. For a particular cost function, there may also be other classes of characters for which the score will be identical, and therefore needs to be computed only once and then multiplied by the number of characters observed in that class. An example will be given below.

The algorithms discussed above give the parsimony score for a single tree. To obtain an estimate of the MP tree, the score of every tree should be computed and the tree or trees with the minimum score should be selected. In practice, however, the number of phylogenies is so large (see Sect. 1.1.1) that it is impractical or impossible to compute a score for every tree once there are more than about eight taxa. Thus, other methods must be used to attempt to estimate the MP tree. This problem of searching for the MP tree is shared by any optimality criterion for phylogenetic inference, and the discussion of methods for identifying optimal phylogenies will be discussed in this general setting in Sect. 1.3. It is not unusual for a search of tree space to result in several trees that share the minimum known parsimony score, as was the case for the cephalopod example considered here. In this case, it is typical to use some kind of consensus method to summarize the main features that are shared by these trees [7].

While the parsimony criterion has been widely applied in phylogenetic studies, there has been some controversy over its desirability as an optimality criterion (see [20]). The idea of finding the most parsimonious explanation of an observed data set is philosophically appealing, but in practice there are situations in which the parsimony method may behave poorly. One problem is that parsimony may not have the property of *consistency*. An estimator of a parameter is said to be consistent if it converges to the true unknown value of the parameter as the amount of data collected increases. In the phylogenetic setting, an estimator of the tree is consistent if it converges to the true tree as more characters are added to the data matrix. Felsenstein [19] presented a scenario in which parsimony leads to inconsistent estimates. Trees satisfying the conditions specified by Felsenstein are said to fall in the *Felsenstein zone*, which will be described later.

Consider a tree with four taxa, for which three unrooted topologies are possible (see Table 1.5). To simplify the computations, suppose that each character can be in one of two states, 0 and 1, and that changes between the two states are weighted equally. There are then $2^4 = 16$ possible character patterns, which can be grouped into five classes based on the similarity of

Table 1.5. Length for each of the 16 possible character patterns on each of the three four-taxon trees

	A–p p–C A⟍ ⟋B A⟍ ⟋B B–q q–D C⟋ ⟍D D⟋ ⟍C		
0011, 1100	0	0	0
0001, 1110 0010, 1101 0100, 1011 1000, 0111	1	1	1
0011, 1100	1	2	2
0101, 1010	2	1	2
0110, 1001	2	2	1

their lengths on the three trees. For example, the patterns 1001 and 0110 will have the same length as one another on all of the trees. Table 1.5 lists the patterns in each class, and the length on each of the three trees.

Let us suppose that the tree in column 2 of Table 1.5 is the true tree, and that the labels on the branches of that tree correspond to the probability of observing different states at the nodes on either end of the branches. Note that p and q must both be less than 0.5, since even for an infinitely long branch, there should be a 50% chance of having the same or a different state at the ends of the branch. We can use the Fitch algorithm to compute the probability of observing any of the character patterns as a function of p and q. Note that when searching for the MP tree, it is not necessary to compute the probability of the set of character patterns in first two rows in Table 1.5, since the score is the same for all three trees in these cases (these sites are not phylogenetically informative). The probabilities for the last three rows in the table are

$$P_{xxyy} = (1-p)(1-q)[q(1-q)(1-p)+q(1-q)p] + pq[(1-q)^2(1-p)+q^2p],$$
$$P_{xyxy} = (1-p)q[q(1-q)p + q(1-q)(1-p)]+p(1-q)[p(1-q)^2+(1-p)q^2],$$
$$P_{xyyx} = (1-p)q[(1-p)q^2+p(1-q)^2]+p(1-q)[q(1-q)p + q(1-q)(1-p)].$$

where $x, y \in \{0, 1\}$ and $x \neq y$.

The MP tree will be the true tree for an infinitely long character matrix (and thus be statistically consistent) whenever the character pattern $xxyy$ has the highest probability. The probabilities above can be compared to examine situations in which this will occur as a function of p and q. After some algebra, it can shown that the condition for this to occur is $q(1-q) > p^2$ [20]. Figure 1.3 shows the region in which the parsimony method is inconsistent. Because this tends to occur when one pair of branches is long relative to the others, this phenomenon has been termed *long branch attraction*. Several studies have debated whether long branch attraction is likely to be a problem in studies involving real data [28, 32, 62], but it is generally accepted that long branches

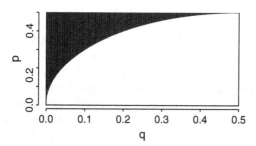

Fig. 1.3. Figure showing the division of the branch length space into zones of consistency and inconsistency. The shaded portion represents the part of the space where parsimony will be inconsistent (i.e., where $q(1-q) < p^2$)

should be subdivided whenever possible to avoid performing estimation in a region of inconsistency.

With a little work, regions of the parameter space where parsimony will be inconsistent can be identified in more general cases. For example, Penny et al. [59] have generalized the above result to allow all of the branch lengths to differ. Felsenstein [20] describes the case where four states, rather than two, are allowed, and points out that the region where inconsistency occurs is smaller when four states are considered. Steel and Penny [74] have shown that parsimony becomes consistent as the number of possible states increases.

In general, parsimony can be expected to perform well when all branches in the tree are relatively short and the problem of long branch attraction can be avoided. Because it is faster than likelihood, it is often preferred in this setting, though in reality it is difficult to be certain a priori that the rate of evolution across the tree has been low. Parsimony shares with any criterion-based method the property that it produces only an estimate of the tree, with no measure of variability in that estimate. It is, therefore, common to apply the bootstrap method (see Sect. 1.4.1) under the parsimony criterion to obtain a sense of how strongly the data support the MP phylogeny.

1.2.2 Maximum Likelihood

The maximum likelihood method was first applied to the problem of phylogenetic inference by Edwards and Cavalli-Sforza (1964) for gene frequency data and by Neyman (1971) for molecular data. Modern techniques for computing the likelihood of a phylogenetic tree were developed by Felsenstein [22], who derived computationally tractable methods for computing likelihoods for trees using molecular data. In this section, we describe the computation of the likelihood of a phylogenetic tree. The likelihood method employs an explicit model for the evolution of molecular data along the branches of a given phylogeny. These models are described in the next section.

Evolutionary Models

The evolutionary models used in phylogenetic inference describe the process of mutation of one nucleotide sequence to another as a function of time. They are often referred to as nucleotide substitution models, since mutation is often thought of as the substitution of one nucleotide for another at a particular site in the DNA sequence. The substitution models most commonly used are Markov models, which have the property that the probability of a nucleotide at a particular site changing from x to y (where both x and $y \in \{A, C, G, T\}$) does not depend on the past history of nucleotides at that site. The probability depends only on the nucleotide x that currently occupies the site and on the time over which the change has occurred. It is also generally assumed that the substitution probabilities do not vary throughout the tree, so that homogeneous Markov processes are used to model sequence change over time, though this can be relaxed [27, 60, 79]. A further common assumption is that of time-reversibility, which means that the process is the same whether it is viewed forward or backward in time. Mathematically, time-reversibility means that $\pi_x P_{xy}(t) = \pi_y P_{yx}(t)$ for all x, y, and t, where $P_{xy}(t)$ is the probability that nucleotide x changes to nucleotide y over time t and π_x is the frequency of nucleotide x at equilibrium.

The most general of the time-reversible models is called the GTR (general time-reversible) model. It is defined by the following instantaneous rate matrix:

$$\mathbf{Q} = \begin{pmatrix} - & \mu a \pi_C & \mu b \pi_G & \mu c \pi_T \\ \mu a \pi_A & - & \mu d \pi_G & \mu e \pi_T \\ \mu b \pi_A & \mu d \pi_C & - & \mu f \pi_T \\ \mu c \pi_A & \mu e \pi_C & \mu f \pi_G & - \end{pmatrix}, \tag{1.5}$$

where the parameter μ is the mean instantaneous rate of mutation and parameters a through f give the rates for the particular type of substitution. The rows and columns of the matrix are ordered A, C, G, and T, so that, for example, the mean instantaneous rate of change of nucleotide A to nucleotide C is $\mu a \pi_C$. The diagonal elements of the $\mathbf{Q}$ matrix are the negative of the sum of the off-diagonal elements in the corresponding rows.

To calculate the likelihood for a tree, we need to use the instantaneous rate matrix $\mathbf{Q}$ to compute the probabilities of changes between the nucleotides. This is done by solving the matrix differential equation

$$\mathbf{P}(t) = e^{\mathbf{Q}t}, \tag{1.6}$$

resulting in a substitution probability matrix with entries $P_{xy}(t)$ that give the probability of nucleotide x changing to nucleotide y over a branch of length t. Note that such a matrix is traditionally referred to as a transition probability matrix, but because the word "transition" has special meaning in the context of nucleotide sequences, the term substitution probability matrix is used in this context instead.

Many submodels of the GTR model are commonly used, the simplest of which is the Jukes–Cantor (JC) model [39]. The JC model assumes that all nucleotides are equally frequent at equilibrium and that the rate of change from any nucleotide to any other is the same, regardless of the nucleotides. This corresponds to setting $\pi_A = \pi_C = \pi_G = \pi_T$ and $a = b = c = d = e = f$ in the $\mathbf{Q}$ matrix above. A slightly more general model is the Kimura two-parameter (K2P) model [40], which allows for a different rate of change for transitions (changes from purine to purine or from pyrimidine to pyrimidine) and for transversions (changes from a purine to pyrimidine or vice versa). This can be accomplished by specifying that $a = c = d = f = 1$ and $b = e = \kappa$, where κ gives the increased rate of change for transitions over transversions, and is motivated by the fact that transitional changes are often observed at higher frequencies. Note that by letting $\kappa = 1$, the K2P model reduces to JC.

Another commonly used model generalizes K2P by allowing the frequencies of the nucleotides to differ. Two common parameterizations of this are given by the HKY85 model [31] and the F84 model [23].

For these submodels of GTR, it is possible to write explicit expressions for the substitution probabilities. For example, for the K2P model, we have

$$P_{xy}(t) = \frac{1}{4} - \frac{1}{4}e^{-\mu t} + \frac{1}{2}(e^{-\mu t} - e^{-\mu t(\kappa+1/2)})\epsilon_{xy} + e^{\mu t(\kappa+1/2)}\delta_{xy}, \qquad (1.7)$$

where δ_{xy} is the indicator function that is 1 when $x = y$ and 0 otherwise, and ϵ_{xy} is 1 if the change from x to y is a transition and 0 otherwise. For the more parameter-rich submodels of GTR, numerical computation of the eigenvalues and eigenvectors of $\mathbf{Q}$ is used to compute the substitution probabilities for particular choices of the parameters a–f. A thorough description of the commonly used evolutionary models is given by Swofford et al. [77].

In applying these substitution models to the estimation of phylogenetic trees under the maximum likelihood criterion, it is often assumed that the substitution rate does not vary across sites. This assumption of rate homogeneity across sites is unrealistic, and can be relaxed in several ways. If the relative rates of substitution across sites were known, then incorporation of these in the substitution probabilities is straightforward. For example, if in the K2P model we let r_h be the rate component at site h, the substitution probability at site h given in (1.7) could be modified by multiplying the term μt by r_h everywhere that it occurs in (1.7).

Alternatively, rate variation over sites could be modeled by some probability distribution (discrete or continuous) with the calculation of sitewise likelihoods (see below) performed by either summing or integrating over this distribution. The gamma distribution is one distribution that has been commonly used to model rate variation across sites [84, 85]. It is also relatively common to assume that some portion of the sites are invariable and do not change among the taxa sampled.

A feature that is common in the data which is not accounted for in the substitution models described above is the presence of *insertions* or *deletions* in the nucleotide sequences. An insertion is the addition of one or more nucleotides to a sequence, while a deletion is the removal of one or more nucleotides from a sequence. Most phylogenetic reconstruction methods in common use either exclude sites which contain an insertion or deletion from the data set or ignore the particular branch leading to an external node with an insertion or deletion in the likelihood calculation for that site. Models that account for insertions and deletions in an appropriate manner have yet to be developed.

Given the number of possible submodels of the GTR model, a practical question arises as to which model should be selected for analysis of a particular data set. It is desirable to use a model that adequately describes the evolutionary processes for the data under consideration, yet parameter-rich models lead to increased computational complexity. A class of methods incorporated into the program ModelTest [63] has recently been proposed to provide selection of the least complex model that is appropriate for a particular data set. The methods work by considering a series of submodels of the GTR model. For each submodel, maximum likelihood estimates of the parameters are obtained using a tree constructed by a distance method. The likelihood score for that tree using the optimized model is computed. Likelihood scores are then compared across a nested class of submodels, and the optimal model is selected by one of three criteria: hierarchical likelihood ratio tests, the Aikaike information criterion (AIC), or the Bayesian information criterion (BIC). This class of methods for model selection in phylogenetics has become enormously popular in the past several years, and is now commonly the first step in performing a maximum likelihood phylogenetic analysis. Recently, however, some cautions in applying the methods have been noted [76]. Because selection of models is clearly an important problem in conducting a phylogenetic analysis, this remains an active area of research.

Markov models of evolution can also be developed for other types of genetic data. For example, for amino acid sequences, the number of possible states increases from 4 to 20, and an instantaneous rate matrix for possible changes between these amino acids can be developed [77]. Because $\mathbf{Q}$ is a 20×20 matrix in this case, it is computationally more difficult to manage. However, two common simplified models are used. The first is a Poisson model which, analogous to the JC model for nucleotide data, specifies equal frequencies for all amino acids at equilibrium as well equal substitution probabilities. The second generalizes the Poisson model by allowing amino acid frequencies to vary. Lewis (2001) proposed a class of Markov models to be used in phylogenetic analysis using likelihood for morphological or other discrete data [44].

Calculation of the Likelihood

Assuming that an appropriate evolutionary model has been selected, the next component of a maximum likelihood analysis is the computation of the likelihood score for a particular phylogenetic tree. Before demonstrating the calculation of the likelihood of a tree, we note some of the properties of the calculation. First, under the assumption of time-reversibility, the calculation of the likelihood does not depend on whether the tree is considered to be rooted or unrooted. Therefore, the first step in most calculations of the likelihood is to place the root at some point in the tree if it is unrooted. Second, it is generally assumed that sites evolve independently of one another. While this is likely to be violated in most real data sets, it simplifies computation tremendously in that it allows the likelihood to be calculated by considering the sites one at a time. The overall likelihood is then the product of the sitewise likelihoods. Formally,

$$l(\tau, \mathbf{t} | D) = \prod_{h=1}^{N} l(\tau, \mathbf{t} | x_h), \qquad (1.8)$$

where $l(\cdot)$ is the likelihood function, τ is the topology, $\mathbf{t}$ is a vector of branch lengths, N is the number of sites in the nucleotide sequences under consideration, X is the data for all of the sites, and x_h is the data for site h only. Usually it is easier to work with the log of the likelihood, so that we are considering

$$L(\tau, \mathbf{t}|\mathbf{X}) = \ln l(\tau, \mathbf{t}|\mathbf{X}) = \sum_{h=1}^{N} \ln l(\tau, \mathbf{t}|x_h). \qquad (1.9)$$

To calculate the likelihood of tree topology τ with branch lengths $\mathbf{t}$ for site h, we must sum the probabilities of all possible scenarios by which the taxa at the tips of the tree could have evolved. Fortunately, Felsenstein [21, 22] provided a simplified version of the calculation, which he termed "pruning" that allows the calculation to be completed without the enumeration of all of these possible scenarios. Let $l_i^h(x)$ be the likelihood for node i at site h, given that node i has state x ($x \in \{A, G, C, T\}$). This is actually the conditional likelihood of the subtree descending from node i, given that node i is in state x. For the tips of the tree, $l_i^h(x)$ will be 1 if the observed nucleotide at site h is x and 0 otherwise. For the internal nodes of the tree, the conditional likelihood will be

$$l_i^h(x) = \sum_{y \in \{A,C,G,T\}} P_{xy}(t_1) l_j^h(y) \sum_{z \in \{A,C,G,T\}} P_{xz}(t_2) l_k^h(z), \qquad (1.10)$$

where y and $z \in \{A, C, G, T\}$, j and k are the descendants of node i, t_1 and t_2 are the lengths of the branches connecting j and i, and k and i, respectively, and $P_{xy}(t)$ is the substitution probability for the change from x to y in time t under the chosen substitution model, as described above. Once the conditional likelihoods for all nodes at all sites have been calculated, the overall log likelihood for the tree is given by

$$\ln l(\tau, \mathbf{t}|X) = \sum_{h=1}^{N} \ln l(\tau, \mathbf{t}|x_h)$$

$$= \sum_{h=1}^{N} \ln \left\{ \sum_{x \in \{A,C,G,T\}} \pi_x l_r^h(x) \right\}, \qquad (1.11)$$

where r is the root node and π_x is the equilibrium frequency of nucleotide x.

While likelihood computations for individual trees can be performed efficiently using Felsenstein's pruning algorithm, the computation is nontrivial in terms of computing time for large data sets. In addition, the likelihood depends on the lengths of the branches in the tree under consideration, and thus branch lengths as well as tree topology must be estimated when searching for the ML tree. Thus, searching for optimal phylogenetic trees under the ML criterion is an important and difficult problem. Current approaches are described in the next section.

1.3 Searching for Optimal Trees

Once an optimality criterion has been selected, the phylogenetic tree must be estimated by finding the particular tree that optimizes that criterion. In this respect, the MP and ML criteria lead to the same difficulty: the optimal phylogenetic tree must be located within a large space of possible phylogenies. For the MP problem, this space consists only of tree topologies, whose number was shown in Sect. 1.1.1 to grow faster than exponentially in the number of taxa. For the ML problem, branch lengths must be considered in addition to topologies, making the search for the optimal tree even more difficult. In this section, we describe some of the common approaches to search for optimal trees under either the MP or the ML criteria. We demonstrate the methods by applying them to the papillomavirus and cephalopod data sets.

1.3.1 Exact Methods

Ideally, all possible trees would be considered, the criterion of interest would be evaluated on each of them, and the tree that maximizes the criterion could be determined exactly. While a relatively simple method can be used to enumerate all possible topologies (Fig. 1.4), it is not computationally possible to evaluate each of them in a reasonable time once the number of taxa exceeds above ten.

Branch-and-bound methods have been developed for this problem, and allow exact searches for up to about 15 taxa for any criterion whose values are monotone as additional taxa are added to the tree. To demonstrate the method, assume that we want to minimize the value of the optimality criterion, which would be the case for MP, for example. Note that for MP the length of the tree cannot decrease as taxa are added to the tree. We begin by specifying an upper bound, C, on the minimum value of the criterion. Referring to Fig. 1.4, we then consider first the four-taxon tree constructed from the three-taxon tree. If the value of the criterion exceeds C, then we need not consider any of the trees that are obtained by adding branches to this four-taxon tree. If the value does not exceed C, then each possible five-taxon tree obtained by adding a branch to the four-taxon tree is considered, and its value of the optimality criterion is compared to C.

The search through the space of trees proceeds in this way. Whenever a tree's value of the optimality criterion exceeds C, it and all trees obtained by adding branches to it are eliminated from consideration, and their values of optimality criterion are never computed. In this way, large portions of the space are never considered, since we can be certain that they do not contain the optimal tree. When this occurs, the algorithm proceeds backward to the largest k-taxon tree whose value did not exceed C. Any $(k + 1)$-taxon trees that have not been previously evaluated are then considered. If at any time in the search an n-taxon tree is encountered with a value C^* of the optimality criteria that is less than C, C is reset to C^*, and the search is continued.

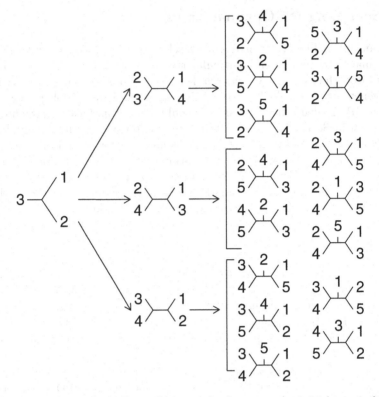

Fig. 1.4. Enumeration of all possible trees for five taxa. An initial tree is formed from the first three taxa (first column). The fourth taxa is added to each possible branch of the tree to obtain the three trees in the middle column. The fifth taxa is then added to each possible branch in each of the four taxon trees to obtain the 15 trees in the third column. These 15 trees are the complete set of possible branching patterns for five taxa. Continuing in this manner would provide an enumeration of all possible trees for any number of taxa

The advantage of using the branch-and-bound method is that it guarantees that a phylogenetic tree that optimizes the criterion of interest is found. Clever programming, including an appropriate choice of the initial upper bound C, can greatly improve the running time of the algorithm [77]. However, the method is limited to data sets that are still rather small in terms of the wealth of sequence data available today. In addition, the method may not necessarily identify trees that are nearly optimal, which might be valuable in a phylogenetic context.

Example: Cephalopod Data

The nine most parsimonious trees used to construct the consensus tree in Fig. 1.1 were found using the branch-and-bound method implemented in

PAUP*. In addition, the molecular data described in Sect. 1.1 can be analyzed under the parsimony criterion using branch-and-bound. For the 28S data, six trees with score 903 were found; the consensus of these trees is shown in Fig. 1.5a. For the histone H3 data, 13 trees with score 271 were found; the consensus of these trees is shown in Fig. 1.5b. The branch-and-bound searches took 15 and 84 s for the 28S and histone H3 genes, respectively. The 18S data is more complex than the data for the other two genes, which is in part due to the much larger number of informative sites for this gene. A branch-and-bound search for this data set is not computationally feasible. The data will be analyzed using heuristic search methods in subsequent sections.

1.3.2 Stepwise Addition and Branch Swapping

The most common method of searching for optimal phylogenies is the method of stepwise addition and branching swapping, implemented in the popular phylogenetic packages PAUP* [78], PHYLIP [25], and fastDNAML [57]. This method begins by constructing an initial tree by successively adding branches to a current tree in the location, which optimizes the value of the criterion. For example, consider the three-taxon topology, for which there is a single internal node and three external edges. A fourth taxon is added to this tree on the edge that gives the best value of the criterion of interest. There are then two internal nodes and five branches. A fifth taxon is then added to the tree on the branch that gives the best value of the optimality criterion. This process continues until a complete tree containing all of the taxa in the data set has been constructed.

A tree constructed using this method, called stepwise addition, is not necessarily optimal, and different orders of addition of taxa can lead to different trees. For this reason, branch swapping is commonly used to search for the optimal tree, using this tree as a starting point. The process of branch swapping consists of rearranging portions of the tree using a specified strategy and evaluating the criterion of interest on the rearranged tree. When the rearranged tree has a better value for the criterion of interest than the current tree, then the rearranged tree replaces the current tree and is subsequently subjected to further branch swapping. The process terminates when no rearrangement of the current tree results in an improvement in the value of the criterion. This tree is then returned as the phylogenetic estimate.

Three rearrangement schemes are commonly used in practice, and these differ in how localized their perturbations of the phylogeny are. The first and most localized move strategy is the nearest neighbor interchange (NNI). NNIs are performed by selecting an internal branch of the tree and then erasing that branch as well as the two branches connecting to each node at the ends of the selected branch (a total of five branches). The result is a set of four subtrees. There are then three ways in which these subtrees can be reconnected to yield a bifurcating topology, two of which differ from the starting topology. Selecting one of these two results in a completed NNI. Because an unrooted

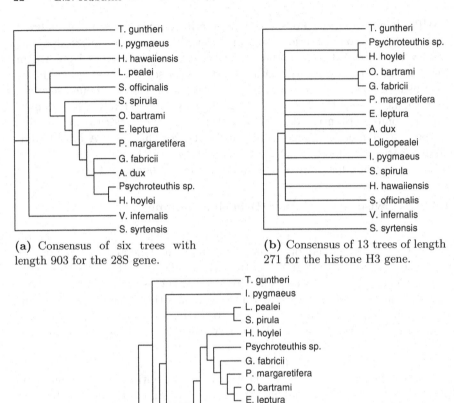

(a) Consensus of six trees with length 903 for the 28S gene.

(b) Consensus of 13 trees of length 271 for the histone H3 gene.

(c) Consensus of two trees with length 4,695 for the 18S gene.

Fig. 1.5. Consensus of collections of most parsimonious trees for the molecular data for the cephalopods. The trees for the 28S and histone H3 genes were obtained using the branch-and-bound search method as implemented in PAUP*. The tree for the 18S data was obtained using 10 random addition TBR searches in PAUP*. The trees shown here are not strictly bifurcating, since some nodes may be unresolved when computing a consensus tree. In addition, the parsimony method does not identify the root of the tree, and hence these trees are unrooted. Note that the Decabrachia are monophyletic for all three genes, and though the relationships within the Decabrachia differ slightly among the trees, the relationship of the Decabrachia to the other taxa is unchanged. Because the outgroup taxon *N. pompilius* was not included in the first two data sets, it cannot be used to root these trees. As a result, the observed relationships for these genes are consistent with both hypotheses described in Fig. 1.1. However, the 18S data set does include *N. pompilius*, and using this to root the tree in (c) provides support for a sister relationship between Vampyromorpha and Decabrachia

topology has $n - 3$ internal nodes and each internal node can produce two NNI moves, $2(n - 3)$ NNI rearrangements are possible.

Subtree pruning and regrafting (SPR) rearrangements are performed by first selecting a branch (either internal or external) and its associated subtree. This subtree is then removed from the tree and reconnected to one of the remaining branches in the tree. It can be shown that selection of an internal branch results in $2n - 8$ possible rearrangements, while selection of an external branch results in $2n - 6$ rearrangements. Because an unrooted bifurcating tree has $n - 3$ internal branches and n external branches, the number of SPR rearrangements is $(n-3)(2n-8)+n(2n-6) = 4(n-2)(n-3)$, though some of these will be the same. Comparing this to the number of NNI rearrangements, we see that SPR moves can result in a more global rearrangement of a given tree.

The most global rearrangement strategy in common use is tree bisection and reconnection (TBR). In TBR rearrangement, an internal branch of the tree is removed and the two resulting subtrees are reconnected in all possible ways. To count the number of possible rearrangements for a particular internal branch, suppose that removal of the selected internal branch divides the tree into subtrees of n_1 and n_2 taxa. There are then $(2n_1-3)(2n_2-3)$ possible ways to reconnect the trees. For a general expression for the number of neighbors of a given tree under TBR moves, see Allen and Steel [4].

The stepwise addition and branch swapping algorithm mentioned above can be modified in several ways. First, branch swapping can be performed at each round of the stepwise addition process, so that the placement of taxa already added to the tree can be updated once more taxa are included. Second, the branch swapping procedure can proceed on all trees that improve the optimality criterion from the current tree, rather than on the first tree found which improves the criterion. While these modifications can increase computational time, possibly substantially, they do provide a more thorough search of tree space.

Even given these modifications, however, stepwise addition and branch swapping comprise a heuristic method with no guarantee that the optimal phylogeny will be found. In practice, the input order of the taxa is randomized and the entire optimization procedure is repeated numerous times from these random starting points, in the hope that this increases the chance of locating the globally optimal tree.

Example: Group A Papillomaviruses

We now consider the papillomavirus data set for which a 1,560-bp portion of the *L1* gene is aligned for 30 sequences. We consider both MP and ML estimation using stepwise addition and branch swapping.

First, we use the program PAUP* [78] to estimate the MP tree. To examine the performance of the stepwise addition and branch swapping method, 20 independent searches using TBR rearrangements were conducted from

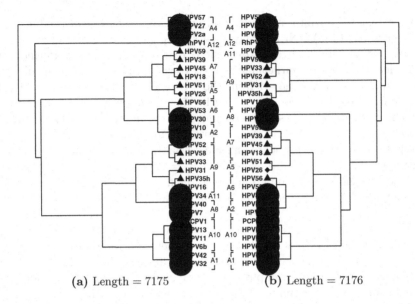

(a) Length = 7175   **(b)** Length = 7176

Fig. 1.6. Two locally optimal trees estimated using parsimony. The tree in (**a**) is believed to be the global optimum

randomly selected starting points. In 17 of the 20 trials, the tree in Fig. 1.6a, which has length 7,175, was found, but on three of the trials, the tree in Fig. 1.6b, which has length 7,176, was found. Note that the two trees differ in their placements of the clade containing Group A2, A5, A6, and A7 as well as the A8 and A9 Groups, and that the trees are not connected to one another using any of the rearrangement strategies discussed in this section. In fact, the symmetric difference [65] between these trees is six, indicating that they are substantially different from one another in terms of topology. Hence they represent local optima or "islands" [50] in the space of trees, and highlight the difficulty of searching for globally optimal trees using criterion-based methods. Although the tree in Fig. 1.6b has a worse value of the optimality criterion, it requires only one more change than the optimal tree and therefore represents a reasonable explanation of the data under the parsimony criterion.

We next consider ML estimation of the papillomavirus phylogeny. Before the phylogeny can be estimated, we must specify an appropriate evolutionary model, including both the selection of a nucleotide substitution model and an assessment of whether the molecular clock should be used. For this example, we use the ModelTest program [63] to select the evolutionary model, and find that, using hierarchical likelihood ratio tests, the GTR + I + G model (GTR model with a portion of the sites invariant and gamma-distributed mutation rates across sites) is preferred. The empirical base frequencies are used to estimate the equilibrium frequencies, resulting in $\hat{\pi}_A = 0.31, \hat{\pi}_C = 0.19, \hat{\pi}_G = 0.20$, and $\hat{\pi}_T = 0.30$. The estimated rate matrix is

$$\hat{\mathbf{Q}} = \begin{pmatrix} - & 4.09 & 7.83 & 2.79 \\ 4.09 & - & 3.46 & 7.06 \\ 7.83 & 3.46 & - & 1.00 \\ 2.79 & 7.06 & 1.00 & - \end{pmatrix}, \tag{1.12}$$

where the four rows and four columns each correspond to nucleotides A, C, G, and T, respectively, and the matrix entries give the estimated rate of change from the nucleotide represented by the row to that represented by the column (see (1.5)).

To consider the appropriateness of the molecular clock, we use the model and parameters above to estimate the phylogeny both with and without the molecular clock assumption. In both cases, the stepwise addition and branch swapping method was used with TBR rearrangements for ten random orderings of the taxa (this is a standard, fairly thorough search for a problem of this size). The null hypothesis in this case is that the data follow a molecular clock (and under this hypothesis, there are $n - 1$ branches in the rooted phylogeny). The alternative hypothesis is that the molecular clock does not hold, and the resulting unrooted phylogeny would then have $2n - 3$ branches. We can perform a likelihood ratio test of the null hypothesis by taking twice the difference in the maximized log likelihoods under the two hypotheses and comparing this to the Chi-square distribution with $2n - 3 - (n - 1) = n - 2$ degrees of freedom. The log likelihood for the ML tree under the molecular clock assumption is $-28,981.26$, and the log likelihood for the ML tree without the clock assumption is $-28,965.56$. The test statistic is then $\lambda = 2(-28,965.56 - (-28,981.26)) = 31.4$. Since λ does not exceed $\chi^2_{28} = 41.3$, we fail to reject the null hypothesis and conclude that these data do not violate the molecular clock assumption. All subsequent analyses described here will be performed assuming that the molecular clock holds.

Using the GTR+I+G model with the parameters estimated above and the molecular clock, PAUP* was used to estimate the ML phylogeny. Twenty independent searches using TBR rearrangements were conducted from random starting points, and the tree shown in Fig. 1.7a was found on every search. Our searches, therefore, did not result in the identification of any local optima with respect to TBR moves among trees. However, ten searches using NNI rearrangements from random starting points did result in the identification of two distinct islands with respect to NNI moves. One island contains the tree identified with the TBR search, which is likely to be the globally optimal tree (Fig. 1.7a). The other tree is shown in Fig. 1.7b and represents a local optima with respect to NNI moves. The two trees differ in their placement of the Group A8, A9, A10, and A11 sequences, and they have a symmetric difference [65] of 4.

Example: Cephalopods

For the cephalopod data set, it was possible to analyze the morphological data and the data for two of the three genes (28S and histone H3) using the

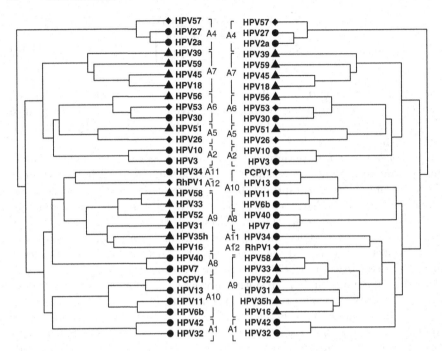

Fig. 1.7. Two locally optimal trees with respect to NNI moves for the papillomavirus data estimated using maximum likelihood. The tree in (**a**) has log likelihood −28,981.26 and is believed to be the global optimum; the tree in (**b**) has log likelihood −28,985.51

branch-and-bound method. However, the 18S molecular data is sufficiently complex that a branch-and-bound search cannot be used. We therefore attempted to find the MP tree for the 18S data using stepwise addition and branch swapping. Ten TBR searches were performed from random starting points and two MP trees were located. The consensus of these trees is shown in Fig. 1.5, where we see that the estimated phylogeny for this gene is similar to those estimated for the 28S and histone H3 genes.

We also consider analysis of the molecular data using maximum likelihood. As a first step, an evolutionary model for each gene is selected using the hLRT implemented in the ModelTest program [63]. For the 18S data, the GTR+G model was selected; for the 28S data, the TrN+G model was selected; and for the histone H3 data, the TrNef+G model was selected (see [63] for complete details of the models). While these three models differ slightly in their complexity, there is agreement in that gamma-distributed rates over sites are necessary, while it is not necessary to model a proportion of the sites as invariant.

Using the ML estimates of the parameters obtained from ModelTest as was done above for the HPV data, ten TBR searches from random starting points were performed for each gene. The resulting trees are shown in Fig. 1.8.

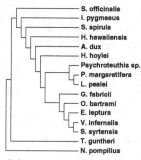

(a) One of two ML trees for the 18S data (log likelihood = −19,711.04).

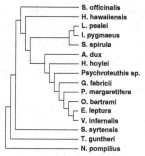

(b) One of two ML trees for the the 18S data (log likelihood = −19,711.17).

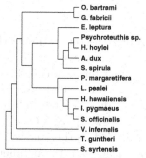

(c) One of two ML trees for the H3 data (log likelihood = −1,647.81).

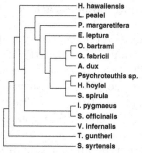

(d) One of two ML trees for the H3 data (log likelihood = −1,647.46).

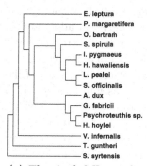

(e) The single ML tree for the 28S data (log likelihood = −4,393.61).

Fig. 1.8. Estimated ML trees for the three molecular data sets in the cephalopod example. The two trees in **(a)** and **(b)** were found to be optimal in ten random addition TBR searches for the 18S data. The two trees in **(c)** and **(d)** were found to be optimal in ten random addition TBR searches for the histone H3 data. The single tree in **(e)** was found to be optimal in ten random addition TBR searches for the 28S data. All of the ML trees agree with a sister relationship between Vampyromorpha and Decabrachia

For the 18S data, the ten searches took a total of 42 min, 36 s, and two trees with very similar scores were obtained. A tree with log likelihood $-19,711.04$ was found on two of the ten searches (Fig. 1.8a), while a tree with log likelihood $-19,711.17$ was found on three of the ten searches (Fig. 1.8b). The ten searches for the histone H3 data took a total of 1 min, 14 s, and again resulted in two trees with very similar log likelihoods: $-1,647.81$ (Fig. 1.8c) and $-1,647.46$ (Fig. 1.8d). For the 28S data, the ten searches took a total of 2 min, 10 s, and a single tree with a log likelihood of $-4,393.61$ (Fig. 1.8e) was found to be optimal. Comparing these trees, we see that they differ primarily in the arrangement within the Decabrachia. All of the trees are consistent with a sister relationship between Vampyromorpha and Decabrachia, in contrast to the results obtained using morphological data (see Fig. 1.1).

Computational Considerations

A few comments on the computational resources used to estimate these phylogenies are warranted. For the HPV data, for example, the average time spent on a single TBR search under the parsimony criterion was 0.039 s, while the average under the ML criterion was 25.6 min for a single NNI search, and 4.07 h for a single TBR search. Empirical studies [57, 69] indicate that the time required by stepwise addition and branch swapping is $O(n^3)$, and so the time increases rapidly as more taxa are included in the analysis. Therefore, searches of this type are not generally practical when the number of taxa is very large (in the hundreds or thousands) for two reasons: (1) the computational time required by a single search becomes prohibitive, and (2) multiple local optima are likely to exist, and so even if these algorithms could be applied to the problem, it is unlikely that they could be applied to enough random starting points in tree space to give an adequate search over the space.

The development of computational methods to handle massive phylogenetic data sets is therefore an area of active research. For example, the program RAxML (Randomly A(x)ccelerated Maximum Likelihood) has recently been applied to data sets containing as many as 25,057 taxa [72, 73]. The program uses stepwise addition and branch swapping with a rearrangement strategy called Lazy Subtree Rearrangement that improves computational efficiency by limiting the scope of SPR moves, as well as many computational modifications designed to improve efficiency. The method is among the most successful methods to be applied to very large phylogenies (>1,000 taxa) to date, the others being PHYML [30] and GARLI [88] (which will be discussed further). However, because it is very difficult to know whether a global maximum has been found, evaluation of the success of such methods for real data of this magnitude remains challenging. The stochastic search methods discussed in the next section provide some alternative techniques that may be promising for large-scale problems.

1.3.3 Stochastic Searches

The problem of finding the optimal phylogeny under either the MP or ML criteria belongs to a class of problems known as NP-complete, meaning that no known polynomial-time algorithm for solving the problem exists ([9, 10]). Algorithms that utilize stochastic, rather than deterministic, searches have been successfully applied to other NP-complete problems [1], and thus many of these stochastic search algorithms have been adapted to the phylogenetic inference problem. Several such algorithms will be described in this section.

The first stochastic algorithm to be applied to the phylogeny estimation problem was the simulated annealing algorithm ([5, 10, 48, 69]). Simulated annealing is a general method for function optimization that is designed to provide an efficient search of tree space while avoiding entrapment in local optima. The algorithm works by making perturbations to a current tree, τ_c, perhaps using one of the rearrangement schemes described in the previous section. The value of the criterion of interest is then evaluated for the perturbed tree. If this tree has a better value of criterion than τ_c, then it becomes the current tree for the next round of perturbation. If value of the criterion is worse, then the decision of whether or not it will replace τ_c is made probabilistically. The probability that the perturbed tree is accepted depends both on the actual value of the criterion as compared with τ_c (with lower probability of acceptance assigned to worse trees) and the number of iterations already performed by the algorithm. As the algorithm proceeds, the probability of accepting trees with worse values of the criterion decreases according to a *cooling schedule* (following the analogy with the physical process of annealing, from which the idea of the algorithm was conceived). Theory underlying the annealing algorithm guarantees that if the search is run long enough and the cooling schedule satisfies certain conditions, the algorithm will converge on the globally optimal solution [1, 2, 49]. In practice, the decision to terminate the algorithm is made when few perturbations of a current tree are being accepted. The simulated annealing algorithm has been successfully applied to both the MP [5] and ML criteria [69] for small- to medium-sized problems. It has not been rigorously tested for very large problems.

Genetic algorithms (GAs) have also been successfully applied to the problem of phylogeny estimation [42, 43, 52, 88]. Like simulated annealing, a genetic algorithm is a general technique for optimization and is based on the analogy of such problems to the genetic transmission of desirable characteristics to successive generations in an evolutionary setting. GAs thus work by considering a population of "solutions" (in this case, phylogenetic trees), each of which is considered a generation in the algorithm. A new population for the next generation is formed from the current generation by applying several mechanisms, all motivated by actual evolutionary processes, to the current generation. For example, some solutions from the current generation will survive to the next generation, based on "good" values of the optimality criteria. Some solutions may be mutated before survival to the next generation. For example, phylogenies can be rearranged according to the schemes described in the previous section before they are transmitted to the next generation. In most implementations, several mechanisms that differ in the scope of the changes they produce are applied in a probabilistic manner to each generation. GAs have been shown to perform well in phylogenetic inference problems, and a recently-developed GA for this problem, called GARLI (Genetic Algorithm for Rapid Likelihood Inference), has been applied to data sets with over 1,000 taxa [88].

A third method of stochastic search that has recently been used for phylogenetic inference is the Ratchet [56, 81]. This method, which was specifically derived for estimating phylogenies, differs from the two methods previously described in that it involves randomly permuting the data, rather than randomly permuting trees. The Ratchet combines traditional hill-climbing methods, such as stepwise addition and branch swapping, with random reweightings of the data to allow the search to more thoroughly explore the space of phylogenetic trees in a computationally efficient way. In particular, the Ratchet alternates between performing a round of branch swapping on the actual data and on a version of the data for which the positions in the DNA sequence have been randomly reweighted, so that some sites are over-represented in the data, while some are under-represented. The idea behind the reweighting is that emphasizing different portions of the data that may possibly contain differing phylogenetic signal may allow the search to jump over broader regions of the space. Once the reweighted data have allowed such a jump to be made, the data is returned to the original weights to determine whether the region of the tree space currently being explored is optimal for the actual data. The Ratchet has been shown to perform well for several data sets of varying size for both MP [56] and ML [81]. An advantage is that short scripts can be written to allow the method to be performed using currently existing software (e.g., see http://www.ucalgary.ca/~dsikes/software2.htm).

Disadvantages of Stochastic Searches

While the potential for stochastic search techniques to improve our ability to locate optimal phylogenies is great, they do have several disadvantages as well. Chief among these is that all of these methods involve setting several parameters that determine how the search will run. For example, applying a simulated annealing algorithm involves setting a cooling schedule; GAs involve specifying probabilities for survival to the next generation and mutations and matings of trees in a given generation; and the Ratchet involves specifying a percentage of characters to reweight, a mechanism for assigning weights, and a determination of how many iterations should be performed. Clearly there is tremendous flexibility in these algorithms for this reason, but systematic study of the properties of the algorithms under varying settings becomes difficult over the range of phylogenetic data sets that might be encountered. Another disadvantage is that the theoretical properties of such algorithms do not necessarily carry over to actual practice, since run time for any given problem cannot be infinite. Perhaps the most promising techniques for the analysis of truly huge data sets (in the hundreds and thousands of taxa) will come from combining the approaches mentioned in this and the last section.

Example: Group A Papillomaviruses

The parsimony ratchet [56] was applied to papillomavirus data set with 200 iterations (each of which consists of a search using the reweighted data and a search using the original data), with 15% of the data randomly reweighted at each iteration. The tree with length 7,175 (see Fig. 1.6a) was found on 159 of these 200 iterations, while the tree with length 7,176 (see Fig. 1.6b) was found on the remaining 41 iterations. While the ratchet performs quite well for this problem, it is especially useful when the number of taxa becomes larger.

1.3.4 Other Heuristic Search Methods

Numerous additional heuristic methods have been developed for the problem of searching for the optimal phylogenetic tree. Notable among these are star decomposition methods and divide-and-conquer methods. Star decomposition methods begin with a single internal node from which n branches lead to each of the external nodes, a so-called "star" phylogeny. At each step of the algorithm, the two nodes that produce the best value of the optimality criterion when paired are joined, until a completely bifurcating tree is obtained. Star decomposition methods can be applied to any optimality criterion that can be evaluated for multifurcating trees. They have been applied to both distance data (leading to the popular neighbor-joining algorithm [68]) and analyses using maximum likelihood [3, 67, 86]. Although fast, these methods often lead to solutions that are only locally optimal. These methods are often effectively used as starting points for other heuristic algorithms, such as branch swapping.

Divide-and-conquer techniques decompose the optimization problem for n taxa into a number of sub-problems each of which consists of inferring the optimal phylogenetic tree for a subset of the taxa, with the idea that the optimal phylogeny is more easily found when the number of taxa is smaller. Once estimates of the phylogeny for each of the subsets are found, the resulting trees are combined to form a "supertree" containing all n of the original taxa. Divide-and-conquer strategies are generally quick, and there is some evidence that their results are reasonably accurate [37]. However, implementation of such strategies is complicated by the need to determine a method for both dividing the data into subsets and recombining those subsets, which becomes particularly difficult in the case of disagreements among the trees estimated for each subset. Among the methods in common use are the Short Quartet Methods [17], Disk-Covering Methods [37], and Quartet Puzzling [75].

1.4 Assessing Uncertainty in Phylogenetic Estimates

Throughout this chapter, the focus has been on obtaining a single estimate of the phylogenetic tree for a collection of n taxa under a particular optimality criterion. In addition to the estimate of the phylogeny, it is desirable to have a measure of confidence in that estimate. For example, how strongly are certain nodes within the phylogeny supported? There are currently two methods available which provide this type of information, the bootstrap and Bayesian phylogenetic analysis. A very brief overview of each of these will be given below, with additional references provided for the interested reader.

1.4.1 The Bootstrap

The bootstrap is a statistical technique that is commonly used to estimate the variance of statistics whose distribution cannot be easily derived [13, 14]. Felsenstein [24] first applied the bootstrap to the phylogenetic inference problem as a method for assessing the support for individual nodes of the estimated phylogeny. Both nonparametric and parametric versions of the bootstrap technique have been applied in a phylogenetic setting. Recall that the data matrix **X** has N columns, x_h, each of which consist of the data for a particular character in the data set (e.g., a site in the DNA sequences). In each replication of a nonparametric bootstrap analysis, these columns are sampled *with replacement* until a new data matrix X^* is produced. The phylogenetic search procedure (including both the optimality criterion and the search method) is then applied to the data matrix X^* to produce an estimated phylogeny. This process is repeated B times, where B is generally at least 100, and a collection of B phylogenies, each estimated from a particular bootstrap sample, is produced. The proportion of times that a particular node appears in the B trees is then referred to as the *bootstrap support* for that node, and is generally assumed to provide a measure of confidence for that node in the data under the particular method of analysis employed. The parametric bootstrap method is similar, except that the B bootstrap data matrices (the X^*) are produced by simulating data along the tree estimated from the original data according to a particular evolutionary model, rather than by resampling the original data.

While the use of the bootstrap is well-accepted in phylogenetic studies, the precise interpretation of the bootstrap support values has been controversial [15, 33]. Nonetheless, the bootstrap is a useful way to assess support for particular nodes in an estimated phylogeny.

1.4.2 Bayesian Methods

In Bayesian analysis, inference concerning a parameter of interest is based on the posterior distribution of that parameter. This posterior distribution is computed by considering both the prior distribution associated with the parameter (which represents the prior beliefs about the value of the parameter before collecting data) and the likelihood of the observed data, computed assuming a particular model. The prior distribution and the likelihood are combined to produce the posterior distribution according to Bayes Theorem.

In the phylogenetic setting, the parameter of interest is the phylogenetic tree, which might include the topology, the branch lengths, and the parameters of the evolutionary model, and we wish to obtain the posterior distribution $P(\text{Tree}|\text{Data})$. Applying Bayes Theorem, we have

$$P(\text{Tree}|\text{Data}) = \frac{P(\text{Data}|\text{Tree}) \times P(\text{Tree})}{P(\text{Data})}, \qquad (1.13)$$

where $P(\text{Data}|\text{Tree})$ is the likelihood function (1.11) and $P(\text{Tree})$ is the prior distribution. Because the unconditional probability $P(\text{Data})$ cannot be easily computed, Markov Chain Monte Carlo (MCMC) is used to approximate the posterior distribution. Once the posterior distribution has been estimated, it can be used to examine any feature that is of interest. Most often, the Bayesian estimate of the phylogeny is taken to be the tree with the largest posterior probability. Support values for each node in the tree are given by the posterior probability associated with that node, which is the frequency with which it occurs in the posterior sample. However, other quantities may be of interest and are readily estimated from the posterior distribution of trees as well.

More detail concerning Bayesian phylogenetic inference is beyond the scope of this chapter, but useful overviews of the topic are given in [34, 36] and a widely-used MCMC method for Bayesian phylogenetic inference is implemented in the program MrBayes [35]. The use of Bayesian techniques in phylogenetic inference has increased rapidly in the past five years, and increasingly sophisticated models and algorithms continue to be developed in this framework.

1.5 Conclusions and a Look to the Future

In this chapter, we have described methods for estimating phylogenetic trees using both morphological and molecular data. We have focused on maximum parsimony and maximum likelihood, two of the most commonly used optimality criteria for phylogenetic estimation. Additionally, algorithms to search the large space of phylogenetic trees for the tree or trees that optimize these criteria have also been discussed.

Note that when referring to the molecular data, the methods examined here have assumed that data for a single gene are being analyzed. Thus these methods provide an estimate of the evolutionary history of only that gene. While it is expected that the histories of individual genes are similar to the overall evolutionary history of the species, these need not agree exactly. Several biological mechanisms can give rise to differences in the histories of individual genes, including hybridization [16, 47], deep coalescence [51, 58, 66], horizontal gene transfer [53, 80], and gene duplication [29]. Methods to model the relationship between gene histories and the overall species phylogeny are just beginning to be developed [12, 70], and will likely be an important focus of future research in this area.

With the advent of DNA sequencing technology that allows for rapid production of molecular data, data sets containing multiple genes or even whole genome data are becoming increasingly common. The challenge, then, is to adapt and expand current phylogenetic methods to handle such genome-scale data. This new field of phylogenomics [61] will provide a rich source of problems in the years to come.

1.6 Web Resources

The data sets and PAUP* code used for the analysis in this chapter are available at Mathematical Biosciences Institute website:
http://www.mbi.osu.edu/2005/tutorials2005.html

Acknowledgement

I thank Drs. Dennis Pearl and Paul Fuerst for helpful discussions on an earlier draft of this chapter, which is based on the MBI tutorial that they presented. Annie Lindgren kindly provided the cephalopod data and gave permission for its inclusion.

References

[1]. Emile Aarts and Jan Korst. *Simulated Annealing and Boltzman Machines.* Wiley and Sons, First edition, 1989.

[2]. Emile Aarts and P. J. VanLaarhoven. A new polynomial time cooling schedule. *Proc. IEEE Int. Conf. On Computer-Aided Design, Santa Clara,* x:206–208, 1989.

[3]. J. Adachi and M. Hasegawa. *MOLPHY: Programs for molecular phylogenetics I - PROTML: Maximum likelihood inference of protein phylogeny.* Computer Science Monographs, No. 27. Institute of Statistical Mathematics, Tokyo, 1992.

[4]. B. L. Allen and M. Steel. Subtree transfer operations and their induced metrics on evolutionary trees. *Annals of Combinatorics,* 5:1–13, 2001.

[5]. Daniel Barker. *LVB 1.0: Reconstructing Evolution with Parsimony and Simulated Annealing.* University of Edinburgh, 1997.

[6]. L. Bonnaud, R. Boucher-Rodoni, and M. Monnerott. Phylogeny of cephalopods inferred from mitochondrial DNA sequences. *Molecular Phylogenetics and Evolution,* 7:44–54, 1997.

[7]. D. Bryant. *A classification of consensus methods for phylogenetics,* pages 163–183. DIMACS Series in Discrete Mathematics and Theoretical Computer Science, Volume 61. American Mathematical Society, 2003.

[8]. D.B. Carlini, R.E. Young, and M.V. Vecchione. A molecular phylogeny of the Octopoda(Mollusca: Cephalopoda) evaluated in light of morphological evidence. *Molecular Phylogenetics and Evolution*, 21:388–397, 2001.

[9]. B. Chor and R. Tuller. Maximum likelihood of evolutionary trees: hardness and approximation. *Bioinformatics*, 21(Suppl. 1):i97–i106, 2005.

[10]. Andreas Dress and Michael Kruger. Parsimonious phylogenetic trees in metric spaces and simulated annealing. *Advances in Applied Mathematics*, 8:8–37, 1987.

[11]. A.W.F. Edwards and L. L. Cavalli-Sforza. *Reconstruction of evolutionary trees*, pages 67–76. Systematics Association Publication No. 6, 1964.

[12]. S.V. Edwards, L. Liu, and D. K. Pearl. High resolution species tree without concatenation. *Proceedings of the National Academy of Sciences USA*, 104:5936–5941, 2007.

[13]. B. Efron. *The Jackknife, the Bootstrap, and Other Resampling Plans*. CBMS-NSF Regional Conference Series in Applied Mathematics, Monograph 38. Society for Industrial and Applied Mathematics, Philadelphia, 1982.

[14]. B. Efron and R. Tibshirani. *An Introduction to the Bootstrap*. Chapman and Hall, New York, 1993.

[15]. Bradley Efron, Elizabeth Halloran, and Susan Holmes. Bootstrap confidence levels for phylogenetic trees. *Proc Natl Acad Sci USA*, 93:13429–13434, 1996.

[16]. N.C. Ellstrand, R. Whitkus, and L. H. Rieseberg. Distribution of spontaneous plant hybrids. *Proceedings of the National Academy of Sciences*, 93(10):5090–5093, 1996.

[17]. P. Erdos, M. Steel, L. Szekely, and T. Warnow. Local quartet splits of a binary tree infer all quartet splits via one dyadic inference rule. *Computers and Artificial Intelligence*, 16(2):217–227, 1997.

[18]. J.S. Farris. Methods for computing Wagner trees. *Systematic Zoology*, 19:83–92, 1970.

[19]. J. Felsenstein. Cases in which parsimony or compatibility methods will be positively misleading. *Systematic Zoology*, 27:401–410, 1978.

[20]. J. Felsenstein. *Inferring Phylogenies*. Sinauer Associates, 2004.

[21]. Joseph Felsenstein. Maximum-likelihood and minimum-steps methods for estimating evolutionary trees from data on discrete characters. *Systematic Zoology*, 22:240–249, 1973.

[22]. Joseph Felsenstein. Evolutionary trees from DNA sequences: a maximum likelihood approach. *Journal of Molecular Evolution*, 17:368–376, 1981.

[23]. Joseph Felsenstein. Distance methods for inferring phylogenies: A justification. *Evolution*, 38:16–24, 1984.

[24]. Joseph Felsenstein. Confidence limits on phylogenies: An approach using the bootstrap. *Evolution*, 39(4):783–791, 1985.

[25]. Joseph Felsenstein. *Phylogenetic Inference Package (PHYLIP), Version 3.5*. University of Washington, Seattle, 1993.

[26]. W.M. Fitch. Toward defining the course of evolution: Minimum change for a specified tree topology. *Systematic Zoology*, 20:406–416, 1971.

[27]. N. Galtier. Maximum likelihood phylogenetic inference under a covarion-like model. *Molecular Biology and Evolution*, 18:866–873, 2001.

[28]. A. Graybeal. Is it better to add taxa or characters to a difficult phylogenetic problem? *Syst. Biol.*, 47:9–17, 1998.

[29]. R. Guigo, I. Muchnik, and T.F. Smith. Reconstruction of ancient molecular phylogeny. *Molecular Phylogenetics and Evolution*, 6:189–213, 1996.

[30]. S. Guindon and O. Gascuel. A simple, fast, and accurate algorithm to estimate large phylogenies by maximum likelihood. *Syst. Biol.*, 52(5):696–704, 2003.

[31]. Masami Hasegawa, Hirohisa Kishino, and Taka-Aki Yano. Dating of the human-ape splitting by a molecular clock of mitochondrial DNA. *Journal of Molecular Evolution*, 21:160–174, 1985.

[32]. S.M. Hedtke, T.M. Townsend, and D.M. Hillis. Resolution of phylogenetic conflict in large data sets by increased taxon sampling. *Syst. Biol.*, 55(3): 522–529, 2006.

[33]. David Hillis and James Bull. An empirical test of bootstrapping as a method for assessing confidence in phylogenetic analysis. *Syst Biol*, 42(2):182–192, 1993.

[34]. J.P. Huelsenbeck and J.P. Bollback. *Application of the likelihood function in phylogenetic analysis*, pages 415–444. Wiley (edited by D. J. Balding, M. Bishop and C. Cannings), 2001.

[35]. J.P. Huelsenbeck and F. Ronquist. MrBayes3: Bayesian phylogenetic inference under mixed models. *Bioinformatics*, 19:1572–1574, 2003.

[36]. J.P. Huelsenbeck, F. Ronquist, R. Nielsen, and J. P. Bollback. Bayesian inference of phylogeny and its impact on evolutionary biology. *Science*, 294(5550):2310–2314, 2001.

[37]. Daniel Huson, Scott Nettles, and Tandy Warnow. Disk-covering, a fast converging method for phylogenetic tree reconstruction. *Journal of Computational Biology*, 6(3):369–386, 1999.

[38]. J.G. Joung, S. June, and B.T. Zhang. Protein sequence-based risk classification for human papillomaviruses. *Computers in Biology and Medicine*, 36(6):656–667, 2006.

[39]. T.H. Jukes and C.R. Cantor. Evolution of protein molecules. In H. N. Munro, editor, *Mammalian Protein Metabolism*, pages 21–132. Academic Press, New York, 1969.

[40]. Motoo Kimura. A simple method for estimating evolutionary rate of base substitutions through comparative studies of nucleotide sequences. *Journal of Molecular Evolution*, 16:111–120, 1980.

[41]. A.G. Kluge and J.S. Farris. Quantitative phyletics and the evolution of anurans. *Systematic Zoology*, 18:1–32, 1969.

[42]. A.R. Lemmon and M.C. Milinkovitch. The metapopulation genetic algorithm: an efficient solution for the problem of large phylogeny estimation. *Proceedings of the National Academy of Sciences*, 99(16):10516–10521, 2002.

[43]. Paul Lewis. A genetic algorithm for maximum-likelihood phylogeny inference using nucleotide sequence data. *Molecular Biology Evolution*, 15(3):277–283, 1998.

[44]. P.O. Lewis. A likelihood approach to estimating phylogeny from discrete morphological character data. *Systematic Biology*, 50:913–925, 2001.

[45]. Wen-Hsiung Li. *Molecular Evolution*. Sinauer Associates, First edition, 1997.

[46]. A.R. Lindgren, G. Giribet, and M.K. Nishiguchi. A combined approach to the phylogeny of cephalopoda (mollusca). *Cladistics*, 20:454–486, 2004.

[47]. T.K. Lowrey, C.J. Quinn, R. K. Taylor, R. Chan, R. Kimball, and J. C. De Nardi. Molecular and morphological reassessment of relationships within the

Vittadinia group of Astereae (Asteraceae). *American Journal of Botany*, 88:1279–1289, 2001.

[48]. M. Lundy. Applications of the annealing algorithm to combinatorial problems in statistics. *Biometrika*, 72(1):191–198, 1985.

[49]. M. Lundy and A. Mees. Convergence of an annealing algorithm. *Mathematical Programming*, 34:111–124, 1986.

[50]. David Maddison. The discovery and importance of multiple islands of most-parsimonious trees. *Systematic Zoology*, 40(3):315–328, 1991.

[51]. Wayne P. Maddison. Gene trees in species trees. *Syst. Biol.*, 46:523–536, 1997.

[52]. Hideo Matsuda. Protein phylogenetic inference using maximum likelihood with a genetic algorithm. In *Pacific Symposium on Biocomputing*, pages 512–523, 1996.

[53]. C. Medigue, T. Rouxel, P. Vigier, A. Henaut, and A. Danchin. Evidence for horizontal gene transfer in *Escherichia coli* speciation. *Journal of Molecular Biology*, 222:851–856, 1991.

[54]. S.B. Needleman and C.D. Wunsch. A general method applicable to the search for similarities in the amino acid sequences of two proteins. *Journal of Molecular Biology*, 48:443–453, 1970.

[55]. M. Nei. *Molecular Population Genetics and Evolution*. North-Holland, First edition, 1975.

[56]. K. Nixon. The parsimony ratchet, a new method for rapid parsimony analysis. *Cladistics*, 15(4):407–414, 1999.

[57]. Gary Olsen, Hideo Matsuda, Ray Hagstrom, and Ross Overbeek. FastDNAml: A tool for construction of phylogenetic trees of DNA sequences using maximum likelihood. *Computations in Applied Biosciences*, 10(1):41–48, 1994.

[58]. Pekka Pamilo and Masatoshi Nei. Relationships between gene trees and species trees. *Molecular Biology and Evolution*, 5(5):568–583, 1988.

[59]. D. Penny, M.D. Hendy, and M.A. Steel. *Testing the theory of descent*, pages 155–183. Oxford University Press (edited by M.M. Miyamoto and J. Cracraft), 1991.

[60]. D. Penny, B.J. McComish, M.A. Charleston, and M.D. Hendy. Mathematical elegance with biochemical realism: The covarion model of molecular evolution. *Journal of Molecular Evolution*, 54:711–723, 2001.

[61]. H. Philippe, F. Delsuc, H. Brinkmann, and N. Larillot. Phylogenomics. *Annual Review of Ecology, Evolution, and Systematics*, 36:541–562, 2005.

[62]. S. Poe and D. L. Swofford. Taxon sampling revisited. *Nature*, 398:299–300, 1999.

[63]. D. Posada and K.A. Crandall. Modeltest: testing the model of DNA substitution. *Bioinformatics*, 14(9):817–818, 1998.

[64]. Chenna R., H. Sugawara, T. Koike, R. Lopez, T.J. Gibson, D.G. Higgins, and J.D. Thompson. Multiple sequence alignment with the clustal series of programs. *Nucleic Acids Research*, 31:3497–3500, 2003.

[65]. D.R. Robinson and L.R. Foulds. Comparison of phylogenetic trees. *Math. Biosci.*, 53:131–147, 1981.

[66]. Noah A. Rosenberg. The probability of topological concordance of gene trees and species trees. *Theor. Popul. Biol.*, 61:225–247, 2002.

[67]. N. Saitou. Maximum likelihood methods. *Meth. Enzymol.*, 183:584–598, 1990.

[68]. N. Saitou and M. Nei. The neighbor-joining method: a new method for reconstructing phylogenetic trees. *Molecular Biology Evolution*, 4:406–425, 1987.

[69]. Laura Salter and Dennis Pearl. A stochastic search strategy for estimation of maximum likelihood phylogenetic trees. *in revision, Systematic Biology,* 2000.

[70]. T. Sang and Y. Zhong. Testing hybridization hypotheses based on incongruent gene trees. *Syst. Biol.,* 49(3):422–434, 2000.

[71]. T.F. Smith and M.S. Waterman. Identification of common molecular sequences. *Journal of Molecular Biology,* 147:195–197, 1981.

[72]. A. Stamatakis. RAxML-VI-HPVC: maximum likelihood-based phylogenetic analyses with thousands of taxa and mixed models. *Bioinformatics - Advanced Access,* 2006.

[73]. A. Stamatakis, T. Ludwig, and H. Meier. RAxML-III: a fast program for maximum likelihood-based inference of large phylogenetic trees. *Bioinformatics,* 21(4):456–463, 2005.

[74]. M.A. Steel and D. Penny. Parsimony, likelihood, and the role of models in phylogenetics. *Molecular Biology and Evolution,* 17:839–850, 2000.

[75]. Korbinian Strimmer and Arndt von Haeseler. Quartet puzzling: A quartet maximum-likelihood method for reconstructing tree topologies. *Molecular Biology and Evolution,* 13(7):964–969, 1996.

[76]. J. Sullivan and P. Joyce. Model selection in phylogenetics. *Ann. Rev. Ecol. Evol. Syst.,* 36:445–466, 2005.

[77]. Dave Swofford, Gary Olsen, Peter Waddell, and David Hillis. *Phylogenetic Inference, in Molecular Systematics (edited by D. Hillis, C. Moritz, and B. Mable),* pages 407–514. Sinauer Associates, Inc., Second edition, 1996.

[78]. David Swofford. *PAUP*. Phylogenetic analysis using parsimony (* and other methods). Version 4.* Sinauer Associates, 1998.

[79]. C. Tuffley and M.A. Steel. Modelling the covarion hypothesis of nucleotide substitution. *Mathematical Biosciences,* 147:63–91, 1998.

[80]. Anna Maria Valdez and Daniel Pinero. Phylogenetic estimation of plasmid exchange in bacteria. *Evolution,* 46(3):641–656, 1992.

[81]. R.A. Vos. Accelerated likelihood surface exploration: the likelihood ratchet. *Syst. Biol.,* 52(3):368–373, 2003.

[82]. M.S. Waterman. *Introduction to Computational Biology.* Chapman & Hall, 1995.

[83]. W.C. Wheeler and D. Gladstein. Malign: A multiple sequence alignment program. *Journal of Heredity,* 85:417–418, 1994.

[84]. Ziheng Yang. Maximum-likelihood estimation of phylogeny from DNA sequences when substitution rates differ over sites. *Molecular Biology Evolution,* 10(6):1396–1401, 1993.

[85]. Ziheng Yang. Maximum likelihood phylogenetic estimation from DNA sequences with variable rates over sites: approximate methods. *Journal of Molecular Evolution,* 39:306–314, 1994.

[86]. Ziheng Yang. *Phylogenetic Analysis by Maximum Likelihood (PAML), Version 1.3.* Department of Integrative Biology, University of California at Berkeley, 1997.

[87]. Z.M. Zheng and C.C. Baker. Papillomavirus genome structure, expression, and post-transcriptional regulation. *Frontiers in Biosciences,* 11:2286–2302, 2006.

[88]. D. Zwickl. Genetic algorithm approaches for the phylogenetic analysis of large biological sequence data sets under the maximum likelihood criterion. Technical report, Ph.D. Dissertation, University of Texas at Austin, 2006.

Large-Scale Phylogenetic Analysis of Emerging Infectious Diseases

D. Janies[1] and D. Pol[1,2]

[1] Department of Biomedical Informatics,
   The Ohio State University, Columbus, OH 43210, USA
   email: Daniel.Janies@osumc.edu
[2] Mathematical Biosciences Institute,
   The Ohio State University, Columbus, OH 43210, USA
   email: dpol@mbi.osu.edu

Summary. Microorganisms that cause infectious diseases present critical issues of national security, public health, and economic welfare. For example, in recent years, highly pathogenic strains of avian influenza have emerged in Asia, spread through Eastern Europe, and threaten to become pandemic. As demonstrated by the coordinated response to Severe Acute Respiratory Syndrome (SARS) and influenza, agents of infectious disease are being addressed via large-scale genomic sequencing. The goal of genomic sequencing projects are to rapidly put large amounts of data in the public domain to accelerate research on disease surveillance, treatment, and prevention. However, our ability to derive information from large comparative genomic datasets lags far behind acquisition. Here we review the computational challenges of comparative genomic analyses, specifically sequence alignment and reconstruction of phylogenetic trees. We present novel analytical results on two important infectious diseases, Severe Acute Respiratory Syndrome (SARS) and influenza.

SARS and influenza have similarities and important differences both as biological and comparative genomic analysis problems. Influenza viruses (Orthymxyoviridae) are RNA based. Current evidence indicates that influenza viruses originate in aquatic birds from wild populations. Influenza has been studied for decades via well-coordinated international efforts. These efforts center on surveillance via antibody characterization of the hemagglutinin (HA) and neuraminidase (N) proteins of the circulating strains to inform vaccine design. However, we still do not have a clear understanding of (1) various transmission pathways such as the role of intermediate hosts like swine and domestic birds and (2) the key mutation and genomic recombination events that underlie periodic pandemics of influenza. In the past 30 years, sequence data from HA and N loci has become an important data type. In the past year, full genomic data has become prominent. These data present exciting opportunities to address unanswered questions in influenza pandemics.

SARS is caused by a previously unrecognized lineage of coronavirus, SARS-CoV, which like influenza has an RNA based genome. Although SARS-CoV is widely believed to have originated in animals, there remains disagreement over the candidate animal source that lead to the original outbreak of SARS. In contrast to the long history of the study of influenza, SARS was only recognized in late 2002 and the virus that causes SARS has been documented primarily by genomic sequencing.

In the past, most studies of influenza were performed on a limited number of isolates and genes suited to a particular problem. Major goals in science today are to understand emerging diseases in broad geographic, environmental, societal, biological, and genomic contexts. Synthesizing diverse information brought together by various researchers is important to find out what can be done to prevent future outbreaks [JON03]. Thus comprehensive means to organize and analyze large amounts of diverse information are critical. For example, the relationships of isolates and patterns of genomic change observed in large datasets might not be consistent with hypotheses formed on partial data. Moreover when researchers rely on partial datasets, they restrict the range of possible discoveries.

Phylogenetics is well suited to the complex task of understanding emerging infectious disease. Phylogenetic analyses can test many hypotheses by comparing diverse isolates collected from various hosts, environments, and points in time and organizing these data into various evolutionary scenarios. The products of a phylogenetic analysis are a graphical tree of ancestor–descendent relationships and an inferred summary of mutations, recombination events, host shifts, geographic, and temporal spread of the viruses. However, this synthesis comes at a price. The cost of computation of phylogenetic analysis expands combinatorially as the number of isolates considered increases. Thus, large datasets like those currently produced are commonly considered intractable. We address this problem with synergistic development of heuristics tree search strategies and parallel computing.

2.1 Introduction

Phylogenetics is the study of the evolutionary relationships of genes and organisms, thus providing a retrospective analysis of biological change and adaptation over time. Phylogenetic trees are represented by acyclic graphs in which the leaves of these graphs represent the observed biological entities (taxa) being compared (e.g., sequences of genes, genomes, and/or anatomy of individuals, isolates or cultivars, species, or any higher level taxonomic unit). The internal nodes of the tree are interpreted as a nested set of hypothetical evolutionary ancestors of the entities under consideration as depicted in Fig. 2.1. Once a tree is complete, changes such as mutations and host shift can be traced along branches of the tree that contain important disease causing strains. This retrospective analysis of features provides means of finding mutations that are diagnostic of pathogens, correlating phenotypes and genotypes, and predicting strains that are important for vaccine design.

2.1.1 Modern School of Phylogenetics

The classification of organisms dates back to Aristotle [ARI343]. However, it was only a few decades ago that the theoretical foundations of the field of phylogenetics as it is practiced today were established.

The modern school of phylogenetics arose from the application of the ideas, termed cladistics, originally proposed by Hennig [HEN66]. Cladistics lead

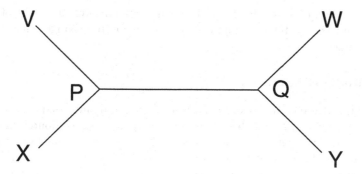

Fig. 2.1. Phylogenetic tree of four taxa labeled V, W, X, and Y and two hypothetical ancestors labeled P and Q

biologists to use shared derived similarities (termed synapomorphies) that distinguish various natural groups of organisms. Nested sets of natural groups of organisms based on synapomorphies are then used to discover the evolutionary relationships between organisms and reconstruct patterns of modification in the features of organisms. Subsequently, these principles have been used to develop optimization techniques to find the most justifiable sets of synapomorphies in large datasets. Optimization techniques are necessary with most large and real world datasets as they often contain several, often conflicting, evolutionary signals (treated below).

In contrast, advocates of another way of thinking, termed phenetics [SNE73], group organisms based on gross measures of similarity. Groups are based on measures evolutionary distance rather than the concept of shared derived characters. In modern practice, similarity methods espousing phenetic concepts are used in searches of nucleotide databases and some multiple alignment methods. Clustering algorithms in which least distant groups are clustered first and then distant clusters are connected are termed distance methods, in phylogenetics. Distance methods typically produce a single tree and cannot, on their own, trace patterns of change in the features of organisms as they convert raw data to distances. Next we discuss how various viewpoints have influenced methods, algorithms, and implementations in phylogenetics.

2.1.2 Phylogenetic Methods

A wide variety of methods have been proposed in order to infer the phylogenetic relationships of organisms. Most methods are based on minimizing edit cost (such as a Hamming distance) to transform one string of nucleotides or organismal characters into another. Phylogenetic methods can be further classified in two different categories: distance-based and character-based. In this paper, we compare and contrast the applications of distance and character-based methods used in infectious disease research. We illustrate applications of these technique to study the evolutionary relationships of groups of RNA viruses and the patterns of mutations and phenotypic changes that can be reconstructed.

Distance-based Methods

Among the distance-based methods, the most commonly used is Neighbor-Joining [SAI87]. Distance based methods require a precomputed multiple alignment of DNA or amino acid sequences drawn from homologous genes. The most similar pair of taxa (as represented by sequences) are clustered. The clustered pair is then considered as a single taxon and the next most similar pair of taxa is clustered until only the last taxon is joined and the tree is completed. Although the use of distance-based methods is relatively common in analysis of organisms that cause infectious disease, several authors have criticized the performance of this method for phylogenetic reconstruction (see [FAR96]). One strategic flaw of the method is that it is computationally greedy. Distance methods form the most similar clusters instantaneously without considering locally suboptimal paths that may lead to a better global optimum.

Character-based Methods

Other methods of phylogenetic analysis focus on characters, which are typically polymorphisms, recognized in columns of aligned nucleotides or amino acids from sequences of interest or investigator encoded characters of polymorphic phenotypes.

Character-based methods seek to find the phylogenetic trees that optimize a particular criterion. Major optimality criteria include parsimony [FAR83] and maximum likelihood [FEL73, FEL81]. Bayesian analysis [RAN96, LI00, HUE02] uses a maximum likelihood optimality criterion but incorporates the probability, termed the prior, that a hypothesis is correct in the absence of data.

The unifying feature of character-based methods is that they examine many randomly generated trees each representing an evolutionary hypothesis of character transformations and organismal relationships. As a character based analysis progresses, edit costs are calculated for transformations that candidate tree imply and optimal trees are stored for further consideration and refinement. The concept of optimality can be associated with cladistics or maximum likelihood but not distance methods. Distance techniques lack a measure of tree quality and means to compare trees.

Cladistics employs parsimony as an optimality criterion. The core concept of cladistics is that the least number of transformations in the data implies the most defensible hypothesis. In cladistics, various edit costs can be applied to different genomic and phenotypic transformations. In the case of weighted parsimony the goal of tree search is to minimize weighted costs.

Under the maximum likelihood criterion the probability of the data, given the tree, calculated with a model for nucleotide or amino acid substitution is optimized. The related technique of Bayesian phylogenetic inference uses maximum likelihood to evaluate trees. Bayesian analysis aims to capture a posterior probability distribution of trees. Typically the results of a Bayesian analysis are displayed not as an optimal tree but rather as the probability that a set of evolutionary relationships is "true," given the prior probabilities, the substitution model, and data.

All character-based methods of molecular phylogenetics [cladistics, maximum likelihood (and related Bayesian methods)] rely on explicit assumptions about ancestral character states to polarize transformations of phenotypes and genotypes that can be reconstructed from data. As an example of such assumptions, in character based analyses is the outgroup criterion (treated below). In contrast, distance based methods do not use an outgroup criterion. Distance based methods do not use the outgroup criterion.

Parsimony

Parsimony is a widely used optimality criterion. This criterion is associated with the concept that simpler explanations provide for more supportable hypotheses. In phylogenetics, the most parsimonious tree(s) is that which implies the minimum number of transformations in sequence and/or phenotypic character states among organisms of interest. The biological justification of this use of parsimony is that descent with modification from a common ancestor is a primary pattern of organismal diversification and the record of transformations can be used to reconstruct that pattern. As such, the tree(s) that minimizes the overall number of independent transformations (convergences or reversals in character state) that are needed to explain the observed data are to be preferred [FAR83]. Recombination and horizontal gene transfer as seen among RNA viruses are violation of the assumption of ancestor to descendent evolution, not parsimony per se. Some novel techniques for discovery and understanding of reassortment and horizontal gene transfer have been developed under the parsimony criterion [WAN05, WHE05].

The parsimony score for a tree is measured based on the number of transformations implied by the tree, known as the *tree length* [FAR70]. The tree length is the sum, over all edges, of the Hamming distances between the labels at the endpoints of the edge [RIC97]. The labels located at the leaves of the tree are the observed characteristics (either genotypic of phenotypic) of the organisms being analyzed. The internal nodes are labeled in order to minimize the tree length of each tree being evaluated.

Given a tree and a matrix of features or aligned sequences for each taxon, the tree length is calculated using the Fitch algorithm [FIT71]. This algorithm works in polynomial time with the amount of data being analyzed (both in the number of characters and taxa). Thus for a sequence alignment of thousands of taxa, each of which is labeled with thousands of nucleotides, the tree length of *a particular tree* can be computed using modern implementations of the Fitch algorithm [GOL03] in fractions of a second.

Inferring Evolutionary Events on a Tree

Given a tree and a data matrix of sequences and features, the parsimony method can pinpoint the branches on which certain evolutionary events are inferred to occur between ancestor or descendent. In an infectious disease context, these events can be a shift by a viral lineage from animal to human host. In the case of standard nucleotide sequence analysis, transformation events include substitution mutations (replacement of a given nucleotide by other) and nucleotide insertions and deletion mutations. Some analyses invoke more complex parsimony models with weighted recombination and horizontal transfer events, as well as differentially weighting certain classes of mutation such as transversions (pyrimidine–purine shifts), transitions (pyrimidine–pyrimidine or purine–purine shifts), or insertion–deletion events [WHE05].

Note that in using the Fitch algorithm to optimize a phylogenetic tree, both the tree length and the branch in which a particular transformation event is inferred to occur can be calculated in unrooted or rooted trees. The results of these calculations are independent of the root chosen for the tree. For example, in an unrooted tree relating four taxa known from their nucleotide sequences (see Fig. 2.2), the Fitch algorithm can be used to identify a specific branch of a tree in which a transformation occurs (e.g., a mutation between nucleotides C and T of the third sequence position occurring in the only internal edge of the tree in Fig. 2.3).

However, the polarity of a transformation event is dependent on how the tree is rooted. Inferring polarity of change requires an external criterion, termed the outgroup.

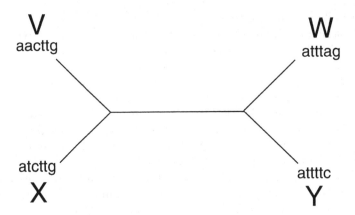

Fig. 2.2. Phylogenetic tree of four taxa V, W, X, and Y as in Fig. 2.1 but with the addition of nucleotide sequences observed for each taxon

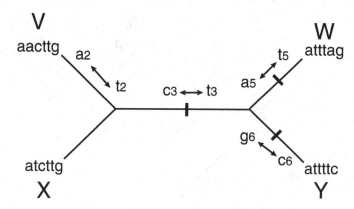

Fig. 2.3. Unrooted phylogenetic tree of four taxa and observed nucleotide sequences as in Fig. 2.2. Here in Fig. 2.3, mutations inferred on various branches are indicated by the nucleotide state and sequence position (in subscript). The number of mutations is four thus the tree length would be four. However, the polarity of mutations cannot be inferred, hence the bidirectional arrows

Polarity of Change and the Outgroup Criterion

In character-based methods explicit assumptions of ancestral character states are set up by the investigator via the designation of at least one taxon as the outgroup. A good outgroup is known to be closely related to the taxa of interest (termed the ingroup). However, the outgroup must be clearly not a member of the ingroup. The underlying logic of the outgroup criterion is that the transformation events that occurred at evolutionary origin of the ingroup can be identified by comparison to modern organisms of another clade but with which the ingroup shares a common ancestor. The common ancestor is a hypothetical organism that provides a baseline set of character states from which polarity determinations can be made. Thus the outgroup method, like Bayesian inference, incorporates some previous knowledge of the relationships of the organisms. If the phylogenetic results show that the ingroup includes some members of the outgroup the previous knowledge must be reevaluated.

The outgroup taxon is included in the data matrix of the phylogenetic analysis and the entire data set is analyzed simultaneously. The phylogenetic position and relationships of the outgroup are determined by the optimality criterion. In the case of the parsimony method, the outgroup is treated as any other taxon and is positioned in the tree in the position that minimized tree length. Once the phylogenetic affinities outgroup are established, the outgroup can be used to root the tree, and the polarities of the transformations can be established (note the unidirectional arrows in Fig. 2.4). If chosen carefully, the outgroup will not be clustered with any of the ingroup. Model based methods can also be used in reconstruction of ancestral character states (e.g., [CHA00, THR04]).

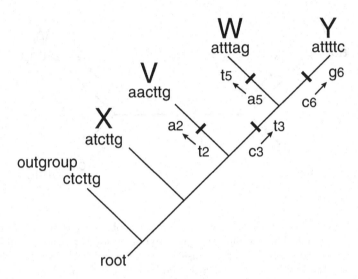

Fig. 2.4. Rooted phylogenetic tree of five taxa. Four of the taxa are the same as in Figs. 2.2 and 2.3 but here we add an outgroup and polarize the mutations, hence the unidirectional arrows. Labels at the leaves of the tree are the observed nucleotide sequences (see Fig. 2.2). Mutations are marked on the branches where they are inferred to occur

To communicate the choice of outgroup taxon or taxa and clarify the relationships of the taxa, character based trees are often drawn as directed acyclic graphs with the root positioned on the branch between the outgroup and ingroup.

In the example diagrammed in Fig. 2.4, a mutation is inferred to occur in the third sequence position from the ancestral state of C to the derived state of T. The presence of a T in the third sequence position is a synapomorphy, a derived character state that can be used to distinguish the members of the group formed by the taxa W and Y. In contrast, the presence of a C in the third sequence position in taxa X and V cannot be used to distinguish these taxa since a C is also present in the third position in the outgroup. In this case the third position C is a primitive similarity of X and Y or a symplesiomorphy. The other mutations occurring in sequence positions 2, 5, and 6 are found only in one taxon and thus cannot be used to infer relationships. These are termed autapomorphies. Sequence position 4 is inferred to have not changed in this example and is thus of no value in discovering groups. In cladistics only the shared derived characteristics, synapomorphies, are used to diagnose a group.

Although other criteria have been proposed to root phylogenetic trees, the outgroup criterion is the least arbitrary. As a result, outgroup rooting is widely used for character-based phylogenetic analyses [NIX94].

Problems of the Outgroup Criterion. As seen, the use of the outgroup taxon provides an informative way test hypotheses on the content of natural groups and to root the phylogenetic tree in a way that allows interpretation of the polarity of change of evolutionary events.

However, the choice of an outgroup taxon is key to the success of this method. If the nucleotide sequences of a candidate outgroup are divergent from the sequences of the ingroup taxa, the phylogenetic position of the outgroup might be hard to establish [WHE90]. Therefore, the choice of the outgroup requires judicious selection and searches for organisms that (1) are safely outside the ingroup but (2) that have comparable data [WHE90].

2.1.3 Sequence Alignment

The cases shown in Figs. 2.1–2.4 are based on the simplifying assumption that the genes sequenced for the taxa of interest have equal number of residues (i.e., amino acid or nucleotide sequences of the same length).

Frequently, in empirical studies of related organisms, homologous genes have sequences with different number of residues. Sequence length variation occurs in both coding and noncoding loci. The causes can be genetic drift, mutation, recombination, or horizontal transfer events. The phylogenetic analysis of molecular sequences, like that of all other comparative data, is based on schemes of putative homology that are then tested via phylogenetic analysis. Unlike some other data types, however, putative homologies in molecular data are not directly observable. Sequences from various organisms are often unequal in length. Hence, the correspondences among sequence positions are not evident and some sort of procedure is required to determine which regions are homologous. This procedure is typically multiple sequence alignment. Alignment inserts gaps to make the putatively corresponding residue line up into columns. These columns (characters) comprise the matrix used to reconstruct cladograms. The matrix is then submitted to phylogenetic analysis in the same manner as other forms of data such as morphological characters scored by an investigator. Thus the primary reason in phylogenetics to create an alignment has a strongly operational basis – to make it possible to submit these data to standard phylogeny programs that were designed to handle column vectors of morphological characters. Nevertheless, alignment followed by tree search is the standard procedure.

Two major options are currently available to analyze sequence data in a phylogenetic framework: a twostep analysis or a one-step analysis.

Two-Step Analyses

Phylogenetic analysis of large genomic datasets can present several nested NP-complete problems: multiple alignment, tree-search, and in some cases, gene order and complement differences among organisms. Just as in distance methods, in most character-based methods, alignments are precomputed before any

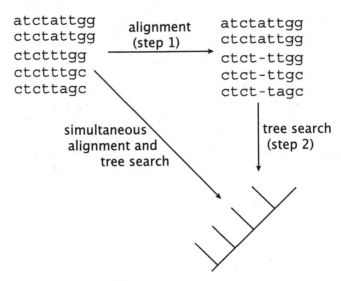

Fig. 2.5. One-step and two-step procedures for the analysis of DNA sequences with different number of nucleotides showing the analysis from raw DNA sequences of different length to the inferred optimal tree

phylogenetic analysis. The alignment procedure is usually done through algorithms that produce a matrix from the raw DNA sequences of the organisms being analyzed (Fig. 2.5). This data set is then analyzed (second step) in order to find the optimal tree (see Sect. 2.1.4).

The multiple alignment procedure ranges from easy in many coding loci to very difficult in noncoding loci such as functional RNAs and genes containing introns [MOR97]. In the case of some protein coding loci the alignment may be a nonissue if there are no significant length differences in sequences. However, various investigators who employ different primer sets and editing styles often produce various length sequences. Leading and trailing gaps produced by experimental artifact should not be counted in tree length calculations.

Results of multiple alignment of functional RNAs and genes containing introns can be sensitive to parameter choices [FIT83]. Important parameters include the addition order of taxa, relative costs of various classes of mutations (transversions, transitions, insertion-deletion), and differential costs applied to opening or extending regions of insertion–deletions. Analyses of different alignments of the same raw sequences can lead to different trees irrespective of tree search procedures [MOR97]. In such cases investigators must search parameter space [WHE95, PHI00] or otherwise justify their assumptions during alignment [GRA03] just as they are required to justify optimality criteria used during tree search.

One-Step Parsimony Analysis

Several researchers have noted that performing phylogenetic analysis into two steps is not consistent with the goals of finding the most parsimonious solutions due to the interdependence of multiple alignment and tree estimation [PHI00, JAN02]. In fact, popular multiple alignment programs such as CLUSTAL [THO94] use a guide tree used to construct the alignment. Therefore, methods have been proposed to make a simultaneous estimation of the optimal sequence alignment and the optimal phylogenetic tree [SAN83]. A modern implementation of the one-step concept in POY [WHE05], termed direct optimization, allows unaligned sequence data to be analyzed without precomputing an alignment. In direct optimization, sequence data are aligned as various trees are built and their optimality is assessed. Thus for each tree considered in a search, various sets of homology statements for the sequence data are considered. One advantage of direct optimization is that the outgroup need not be designated by the investigator. POY allows for randomization of the outgroup taxon and thus adds rigor to the search for optimal trees and homology statements. In some implementations of character-based methods where prealignment is necessary the outgroup can be randomized by scripting a series of analyses, e.g., TNT [GOL03]. One important difference is that in molecular data a rigorous tree search with on a prealigned dataset with unordered characters should lead to the same tree length irrespective of outgroup choice; whereas in direct optimization the homology statements and hence tree length can be dependent on outgroup choice.

Several groups are developing algorithms for simultaneous estimation of alignment and phylogenetic trees. Methods for a one-step phylogenetic analysis have been developed using maximum likelihood [TKF92, FLE05] and parsimony optimality criteria [WHE96], as well as for Bayesian analysis [RED05].

Although the one-step approach has the appeal of using a unified and epistemologically consistent method of alignment and tree estimation, the time and space requirements for computation are considerable. This problem of tree-based alignment is known to be NP-complete [WAN94]. In this situation, genes that vary in length (as most noncoding and intronic containing genes do) present a huge number of possible hypothetical ancestral sequences even for a single binary tree. During a phylogenetic analysis many trees will be examined and compared. For s taxa and l nucleotides per taxon, the cost of computation per tree ranges from $(s-l)l^2$ to $(s-2)l^3$, depending on the heuristics applied. The memory requirements scale proportional to l^3. Fortunately, procedures such as the optimized diagonal transition algorithms described by Ukkonen[UKK85] abate the space and time dependence on l, the number of nucleotides [WHE05].

2.1.4 Tree Searches

Phylogenetic analysis under the parsimony criterion is based on an objective function (tree length). Tree length is used to evaluate the optimality of each phylogenetic tree considered. However, finding the optimal phylogenetic tree (among all possible topologies) is an NP-hard problem [FG82], that resembles the Steiner tree problem. The combinatorial optimization problem of phylogenetic analysis consists of finding the optimal solution from a very large number of possible trees. The number of possible trees increases dramatically with the number of organisms being analyzed [FEL78]. The number of possible (unrooted) phylogenetic trees (T) for a given set of organisms increases following

$$T = (2 \times s - 5)!!, \tag{2.1}$$

where s is the number of organisms (leaves) of the phylogenetic tree. Therefore, the number of possible phylogenetic trees is extremely large even for trees with moderate number of organisms.

As stated above, a phylogenetic analysis consists evaluating topologies in order to find the optimal solution [i.e., the tree(s) with the minimum length]. It is interesting to note that the computing time of an exhaustive evaluation of all possible trees for a fixed number of taxa will increase nearly linearly with the number of characters (e.g. length of DNA sequences) because the Fitch algorithm [FIT71] for evaluating the tree length of a particular topology works in polynomial time.

However, the excessively large number of possible phylogenetic trees of 20 or more organisms (see Table 2.1) makes exhaustive evaluation of all phylogenetic trees intractable.

Note on Multiple Optimal Trees. Frequently, in phylogenetic analysis based on an optimality criterion, there are multiple trees that score the same minimum

Table 2.1. Number of possible unrooted trees as a function of the number of organisms

Organisms	Trees
4	3
5	15
6	105
7	945
10	10^6
20	10^{20}
50	10^{74}
100	10^{182}
1,000	10^{2860}

Order of magnitude is given for the number of trees with more than 10 organisms

for tree length or likelihood. In the set of known optimal trees, the transformations may be differentially distributed and different organismal groups may be implied. Therefore, this set of known optimal trees must be considered equally valuable. These cases represent alternative hypotheses (i.e., phylogenetic trees) that are equally supported by the available data and can be summarized through a *consensus tree*. Several kinds of techniques for consensus estimation exist. The *strict consensus* tree is one of the most frequently used. A strict consensus calculation represents a tree that has all the edges shared by all the known optimal trees. See [SWO91] for further information on various consensus trees.

Exhaustive Searches

Two algorithms can be applied to perform exhaustive searches that evaluate (explicitly or implicitly) all possible phylogenetic trees in order to find the optimal tree. The first of these is exhaustive enumeration, which computes the optimality value (e.g., tree length) of every possible phylogenetic tree and select the tree (or trees) with the minimal value.

The second method is the branch and bound algorithm [HEN82] that implicitly evaluates all possible trees but avoids, in practice, computing all possible trees (see [SEA96]). In current phylogenetic software packages (e.g., [SWO02, GOL03]) this algorithm can be applied to data sets of up to 20 (or 25) organisms and guarantees to find the optimal trees (or trees) for a given phylogenetic data matrix.

Heuristic Searches

For analysis of data sets with larger number of organisms, the number of trees is prohibitively large for conducting an exhaustive search. In modern biology, interesting data sets consider hundreds to thousands organisms. Thus the problem of phylogenetic tree search is compute bound and must be approached through heuristic searches. In these tree searches, a large number of phylogenetic trees are evaluated and the best solution is kept as known estimate of minimum tree length.

Some early examples of heuristic tree searches include the algorithm to compute Wagner Trees [FAR70]. The Wagner algorithm creates a phylogenetic tree of three taxa and progressively adds organisms, attaching them to the branch that generates the minimal increase in tree length at that step. This stepwise procedure is conducted until the last organism is added to the tree. Although this procedure usually results in a tree that has a suboptimal tree length, in most cases this score is significantly better than that obtained with a random choice among all possible topologies. As various starting points are used for building Wagner trees, this aspect of phylogenetic analysis can be considered a type of Monte Carlo randomization.

Given one or many Wagner trees, the next standard heuristic refinement techniques that would be typically applied in tree search are known as branch swapping or hill-climbing procedures (see [SEA96]). This class of refinement procedures consists of performing minor rearrangements of branches in the starting tree. Each Wagner tree is modified by pruning a subtree and reattaching it to a different branch of the remaining tree. The tree length of the modified tree is then calculated. If the modified tree has a shorter tree length it is kept in a buffer of new candidate trees. Branch swapping is applied to all Wagner and candidate trees until the algorithm converges. When no further rearrangements can improve the current topology the branch swapping is finished.

Tree Search Strategies

The results of the branch-swapping algorithm depend on the quality of the starting point (i.e., Wagner tree). In many cases, the tree resulting from the application of branch swapping to a Wagner tree is a local optimum that cannot be further improved by swapping. Therefore, multiple replicates (100s to 1,000s) of independent Wagner trees followed by swapping are typically preformed. At the end of these stages of analysis, the best trees found in all the replicates are kept as a set representing topologies at the known minimum length.

Replication of Wagner builds plus swapping (or random-restart hill climbing) is the most widely used routine implemented in most software packages (e.g., [SWO02, GOL03]). One major drawback of Wagner builds plus swapping is that this procedure is subject to finding only local optima. Finding the globally optimal tree(s) for a dataset of >20 taxa is a NP-hard problem [FG82]. However, performing multiple replicates of this procedure can provide a relative degree of confidence if the minimum length tree(s) converge at the same tree length from numerous independent starting points [GOL99].

Replication of Wagner builds plus swapping is usually efficient for data sets smaller than a hundred taxa. Because of advances in automated DNA sequencing technology, the size of modern comparative data sets far exceed the limits for which these techniques are efficient analytical tools for phylogenetic analysis.

2.1.5 Computational Problems

Large phylogenetic problems are becoming increasingly common across the life sciences due to the prevalence of high throughput nucleotide sequencing technology. Large data sets are of interest to biologists because they provide a rich context of phenotypes and genotypes and permit worldwide and longitudinal sampling of genomes. These large phylogenetic problems will become increasingly common in the years ahead. Thus, phylogenetic methods suited to large datasets will have important consequences not only for the study of organismal classification and evolution, but also for many aspects of public health (see Sect. 2.2). Furthermore, in a operational context, strong organismal sampling has been shown to correlate with improved performance of phylogenetic methods [HIL96, POE98, RAN98, ZWI02, HIL03].

The dauntingly high cost of computation of large-scale phylogenetic analysis stunted this line of research. However, in recent years two main lines of research have provided efficient tools to analyze large phylogenetic datasets: the development of new algorithms and the use of parallel computing.

New Algorithms

Several researchers have combined groups of algorithms into heuristic tree search strategies that have proven to be efficient for phylogenetic analyses of hundreds to thousands of organisms under the parsimony criterion. These heuristic search strategies are based on basic Monte Carlo and hill climbing techniques with the addition of other classes of algorithms including simulated annealing [GOL99], data perturbation [NIX99], divide-and-conquer [GOL99, ROS04], and genetic algorithms [MOI99, GOL99]. Similar search strategies that combine several layers of algorithms have been employed using other optimality criteria such as maximum likelihood [LEW98, SAL01, LEM02, BRA02].

The judicious application of various algorithms has provided efficient solutions for the analysis of datasets of several hundreds organisms [GOL99] in a single CPU [TEH03]. In particular, the successive combination hill-climbing, genetic, and simulated annealing algorithms of tree search have produced a drastic speed up in comparison to other strategies [GOL99]. Efficient implementations of these algorithms have become recently available in software packages [GOL03].

Parallel Computing

The need of phylogenies depicting the evolutionary relationships of datasets consisting of thousands of taxa has prompted the synergistic implementation of efficient heuristic tree search strategies and parallel computing hardware. An increasing number of researchers are developing software suited for parallel computing using Beowulf class clusters [STE00]. Beowulf clusters are simply arrays of commodity PCs and switches enabled by scalable, open source operating systems (e.g. LINUX) and message passing software (e.g. PVM or MPI). Although the advantages of parallel computing in phylogenetics and multiple alignment have been clear for some time [WHE94], the means to exploit this potential for research gain have not been broadly and economically available until the Beowulf concept was developed by the end of the 1990s.

Alignment and tree search problems are naturally suitable for parallel computing. Phylogenetic researchers quickly realized the opportunity presented by Beowulf computing [CER98, JAN01]. Finding an optimal phylogeny requires the evaluation of the same objective function on a large number of alternative trees. Because many trees can be examined concurrently and independently, this has led several authors to implement phylogenetic tree searches in parallel [JON95, SNE00, CHA01, GOL02, BRA02, STA02]. These implementations use the parsimony and maximum likelihood optimality criteria as well as Bayesian analysis. Researchers have also used parallelism to speedup one-step phylogenetic analysis [JAN01] and multiple alignment[WHE94, LI03].

2.2 Applied Phylogenetics

Originally, phylogenetics was considered relevant only to taxonomic and evolutionary studies. However, the ability to identify conserved and divergent regions of genomes is becoming critical data for numerous disciplines in biology and medicine. These fields include vascular genomics [RUB03], ecology [SIL97], physiology [CAR94], pharmacology [SEA03], epidemiology [ROS02], developmental biology [WHI03], and forensics [BUD03]. Phylogenetics has even been used in successful criminal prosecution of a doctor who attempted to cause HIV infection in his former girlfriend via blood products taken from a HIV patient under his care [MET02].

Here we focus on cases in which phylogenetic analyses have helped researchers to understand the evolution and spread of infectious diseases. We provide exemplar cases in which phylogenetic analyses of viral genomes have been crucial to understand complex patterns of transmission among animal and human hosts: Severe Acute Respiratory Syndrome (SARS) [KSI03] and influenza [WEB92].

2.2.1 Phylogenetic Analysis in the Context of Emerging Infectious Disease Research

Emergent infectious diseases often evolve via zoonosis; shifts of an animal pathogen to human host. In fact, most category A pathogens and potential agents of bioterrorism and more than 75% of emergent diseases have zoonotic origins [TAY01, FRA02].

A typical set of tests for the hypothesis of animal host of a disease might be (1) experimentally exposing the candidate host animals with isolated viruses and ascertaining whether infection and viral shedding occurs [MAR04]; or (2) survey populations of animals with antibodies for exposure to the virus [GUA03]. These activities often provide model organisms for vaccine and drug development, data on seroprevalance, and sequence data for viruses isolated from various candidate hosts.

Phylogenetic analysis of genomes is a complement to laboratory and survey studies with a distinct advantage. With phylogenetics, the researcher is not restricted to testing a single hypothesis for a specific candidate host in each experiment. Provided with sequence data for a diverse set of candidate hosts, a researcher performing a single phylogenetic analysis makes a vast number of comparisons, thus evaluating simultaneously many alternative hypotheses. These hypotheses include the evaluation of pathways of transmission among several hosts and the polarity of the transmission events. For example, experimentalists report that small carnivores in Chinese markets have been exposed to SARS-CoV [GUA03] and the virus can infect domestic cats [MAR04]. On their own these data do not necessarily reconstruct the history of the zoonotic and genomic events that underlie the SARS epidemic.

Furthermore, whether or not interspecies transmission is observed or enhanced under controlled laboratory conditions, phylogenetic research is distinct as it can address whether the genomic record has evidence to support a hypothesis for a particular transmission pathway. For example, if phylogenetic analysis reveals multiple independent events of human to avian transmission of influenza viruses without intermediate hosts such as swine that provides a strong argument to reevaluate the hypothesis that pigs serve as "mixing vessels" for avian and human viruses leading to influenza epidemics [SCH90].

In many cases, host shifts occur via recombination between two ancestral pathogen genomes to produce a chimeric descendent. Epidemics can occur when, subsequent to recombination, a lineage of pathogens establishes itself in a new population of hosts, vectors, or reservoir species that can amplify and distribute the pathogen [MOR95]. A host shift can require key mutations and rearrangement of the pathogen genome to infect cells of new hosts followed by adaptation to novel regulatory machinery. Phylogenetics can reconstruct genomic changes at the level of each nucleotide and unravel parental and descendent strains in recombination mediated host shifts [WAN05].

2.3 Evolution of Influenza

Influenza is a widespread respiratory disease caused by an RNA virus (Orthomyxoviridae). The influenza virus has been traditionally divided in three major types: A, B, and C. Influenza viruses of type A are known from many strains that infect both mammal and avian hosts, whereas the other two type are primarily known from humans. Influenza A is characterized by antigenic subtypes (see Sect. 2.3).

Influenza is interesting from both epidemiological and evolutionary points of view due to the interplay between genetic changes in the viral population and the immune system of hosts [EAR02]. There are two basic hypotheses on how influenza A viruses escape the immune response in host population to cause epidemics: (1) antigenic drift, meaning that random point mutations produces novel influenza strains that succeed and persist if they can infect and spread among hosts; (2) antigenic shift, meaning that genes derived from two or more influenza strains reassort thus creating a novel descendent genome with a constellation of genes that can infect and spread among hosts. In both scenarios zoonosis is often involved. In case of antigenic shift the ancestry of only a fraction of the influenza genes may be zoonotic.

Two major classes of influenza epidemics are recognized in humans: seasonal outbreaks and large-scale epidemics known as pandemics [WEB92]. Seasonal influenza is a significant public health concern causing 36,000 deaths and 200,000 hospitalizations in the United States in an average year [GER05]. Elderly and children account for many of these severe cases of seasonal influenza. Much of the population has partial immunity to seasonal influenza strains that are typically descendents of strains circulating in previous years. Pandemics are often caused by infection, replication, and transmission among the human population with influenza strains of zoonotic origin to which few people have prior immunity. Pandemics are rare but can affect the entire human population, irrespective of an individual's predisposition to respiratory diseases. In fact, the 1918 pandemic disproportionately affected young adults [TAU06], suggesting that older adults may have had some immunity.

There have been three major influenza A pandemics, 1918 (H1N1), 1957 (H2N2), and 1968 (H3N2). The pandemic of 1918 is estimated to have killed tens to a hundred million people worldwide and 675,000 in the United States [TAU01]. The Asian flu pandemic of 1957 and the Hong Kong flu pandemic of 1968 were less severe, but caused tens of thousands of deaths in the United States [HHS04]

All of these pandemic strains are thought to have originated in wild birds [WEB92]. The 1957 and 1968 strains are believed to be the results of antigenic shift. However, recent studies suggest that the H1N1 influenza virus that caused the pandemic of 1918 was entirely of avian origin rather than a human-avian reassortant [TAU05]. Other researchers have countered that the 1918 H1N1 strains had a more commonly accepted route to infection of human populations by reassortment in mammals [GIB06; ANT06].

Pandemics can theoretically occur with any strain of influenza. Most influenza infections since 1968 have been attributed to influenza A H3N2 or H1N1 strains. However, there have been several recent reports of novel human infections from avian strains of influenza with subtypes thought to occur rarely in humans. Several cases of human infection of viruses of subtype H7 of avian origin have recently occurred in Canada [TWE04] and the Netherlands [KOO04]. Avian influenza of antigenic subtype H5 and H7 viruses can be found as low or high pathogenic forms depending on the severity of the illness they cause in poultry. Thus far, influenza H9 virus has only been identified as strains with low pathogenicity [LIN00].

Alarmingly, highly pathogenic strains of influenza A with an H5N1 subtype have spread rapidly among various species of birds in China, Southeast Asia, Russia, India, the Middle East, Africa, Eastern, and Western Europe [WHO07a]. These H5N1 influenza A strains share common ancestry with the outbreak of H5N1 that lead to a massive chicken cull and six human deaths in Hong Kong in 1997 [LI04]. Between 2003 and September 10, 2007, there have been 328 cases and 200 deaths among humans [WHO07b]. There are several instances of H5N1 infection of felids and swine in Asia. There is scant evidence of human-to-human transmission in Thailand [UNG05] and Indonesia [YAN 2007]. If lethality to human cases of H5N1 drops, the virus might spread rapidly and without being detected.

Many predict an upcoming avian influenza pandemic of devastating human and economic costs. In the United States alone, it is projected that 15–35% of the population will be affected and the costs could range from 71.6 to 166.5 billion United States (US) dollars [GER05]. Although vaccine production can in theory be modified to include H5N1 strains [DUT05], the genomes of interest are moving targets. It remains unknown whether the descendents of the contemporary H5N1 virus will achieve efficient human-to-human transmission and if this will occur via incremental mutations or a more punctuated reassortment mediated change. Thus phylogenetics is a key technology to track the evolution of H5N1 and compare those changes to genomic and zoonotic events that underlie pandemics.

Subtypes of Influenza Type A

The viruses of influenza type A are classified as various subtypes that represent differences in the antigenic reaction of two key glycoproteins: hemagglutinin (HA) and neuraminidase (NA). These proteins reside on the surface of the virion. These proteins play key roles in recognition and infection of susceptible hosts (HA) and viral replication (NA). These surface proteins are primary antigens recognized by the host immune system [WEB92].

The subtypes of influenza A are labeled according to the reaction of standard monoclonal antibodies to these HA and NA proteins provided by the US Centers for Disease Control to laboratories participating in the World Health Organization's (WHO) surveillance program [HHSb].

Although this number will soon expand, there are currently 16 different antigenic subtypes recognized for HA (labeled from H1 to H16) and 9 different antigenic subtypes of NA (from N1 to N9). Thus, a subtype of influenza virus type A is labeled with the number associated with HA and NA proteins (e.g., the most common subtype found in humans H3N2).

Since 1948, influenza viruses have been the focus of a coordinated surveillance program organized by the WHO [WHO05]. The hemagglutinin gene (HA) is the major target of the influenza surveillance. This program helps track predominant strains to inform the development of new vaccines. Influenza viruses are sampled worldwide through the National Influenza Centers located in 54 countries [WHO05]. Many of the viral isolates sampled by these programs are sequenced for the hemagglutinin gene, although there has been an increasing interest in sampling complete influenza genomes [GHE05, OBE06].

An extensive record of hemagglutinin sequences of the influenza viruses type A isolated since 1902 are publicly available. These data provide a unique set of challenges and opportunities for phylogenetics. The geographically wide and temporally long sampling of viral isolates provides an unprecedented opportunity to study evolutionary patterns underlying the spread and host range of an infectious disease. However, as described earlier, large datasets present an enormous search space of possible evolutionary scenarios to be evaluated.

2.3.1 Phylogenetic Approaches to Influenza Type A

The availability of nucleotide sequences of influenza viruses has triggered numerous research groups to attempt reconstruction of the phylogenetic history of these viruses (e.g., [BUS99, YUA02, FER03, BUS04]). These groups draw on data from currently circulating strains as well as from historically important strains gathered from archival tissue samples. Examples of archival tissues that have provided date of interest to the 1918 epidemic include lung biopsies of deceased soldiers, victims frozen in Alaskan permafrost[TAU97], and waterfowl collected for the Smithsonian in 1916–1917 [FAN02].

Phylogenetic analysis of seasonal influenza sequence data has been used to classify nucleotide substitution mutations. In many codons of the HA gene mutations that produce a change in protein sequence are more frequent than those that do not [BU99]. This finding indicates that selective pressures imposed by the immune system of the hosts can drive the evolution of some codons of HA. Thus an evolutionary perspective can illuminate functional studies of infectious disease [EAR02].

2.3.2 Large Scale Phylogenetic Analysis of HA

As noted, phylogenetics have been widely used to understand history of influenza epidemics, host shifts, as well as evolutionary interactions with the hosts immune system (see Sect. 2.3.1). However, most phylogenetic analyses of influenza thus far have used only fractions of the dataset of influenza nucleotide sequences in the public domain. The sequences in the public domain are largely HA, but recently whole genomes have been produced. The Institute for Genomic Research (TIGR) is rapidly sequencing and releasing into the public domain thousands of influenza genomes under the Microbial Sequencing Center (MSC) program sponsored by the National Institute of Allergy and Infectious Disease (NIAID) [GHE05]. St. Jude Children's Research Hospital in Memphis has contributed a significant increase in the number of avian influenza genomes sequences [OBE06].

Most existing phylogenetic analyses of influenza have focused on the phylogenetic relationships of particular subgroups of influenza type A, such as the H5N1 subtype (e.g., [LI04]) or the H3N2 subtype (e.g., [BUS99]). These analysis have provided useful information but have depicted a disjoint picture of the evolution of the major lineages of influenza. In contrast, other studies have attempted broader subtype scope; however, they included a single viral isolate as an exemplar of each subtype [SUZ02]. This study failed to include an extensive sampling of strains. Poor strain sampling can have a negative impact on the performance of phylogenetic methods (see Sect. 2.1.5) and does not test whether the subtypes are natural groups (i.e. monophyletic). A very recent study has used whole genomes of 136 isolates drawn from a variety of avian influenza subtypes [OBE06].

Here we show results of a comprehensive phylogenetic analysis based on hemagglutinin DNA sequences of 2,359 viral isolates. These sequences include representatives of the 16 different subtypes of the hemagglutinin protein of influenza type A, recorded worldwide by the World Health Organization surveillance program. The analyzed viruses were isolated as early as 1902, from tissues of patients who died during the 1918 Spanish flu epidemic, to recently sequenced isolates from the 2004 seasonal flu and H5N1 outbreak. The analyzed DNA sequences also implies a broad range of host organisms, including multiple species of wild and domestic birds, humans, swine, horses, felids, and whales.

An inclusive phylogenetic analysis with a large number of taxa require the use of efficient tree search strategies (see Sect. 2.1.5) and the use of multiple computers dedicated to the phylogenetic analysis. The cost of computation is tied primarily to the number of strains, not nucleotides. Thus the inclusion of whole genomes does not contribute significantly to the compute bound nature of phylogenetic analysis. However, the inclusion of whole genome data does increase memory demands.

This 2,359 isolate dataset was analyzed with a parallelization of the tree search strategy implemented in a recently developed software for parsimony analysis [GOL03]. The results of this analysis are used here to illustrate two new uses, longitudinal analyses of patterns of zoonotic transmission and assessment of surveillance quality.

Relationships of HA Subtypes

Our results on the relationships of HA subtypes shown in Fig. (2.6) has similarities with the results of Suzuki and Nei [SUZ02], including the clades ((H8 H12) H9), ((H15 H7) H10), ((H4 H14) H3), and (((H2 H5) H1) H6). However, the position of H13 and H11 differ in our trees due to our inclusion of H16. Moreover, the relationship of these clades to one another differs in our assessments. Our tree has a staircase shape with ((H8 H12) H9) basal most, whereas Suzuki and Nei's [SUZ02] tree has a symmetrical shape with no clear basal group.

Host Shifts in HA

Influenza A viruses from wild aquatic birds have been identified as the source of influenza viruses isolated from birds of the order Galliformes (e.g., turkeys, grouse, quails, pheasants, domestic chickens, and their ancestral stock the jungle fowl) [WEB92]. Direct human infection by avian strains of influenza A is considered rare [LIP04]. After the discovery of receptors for both avian

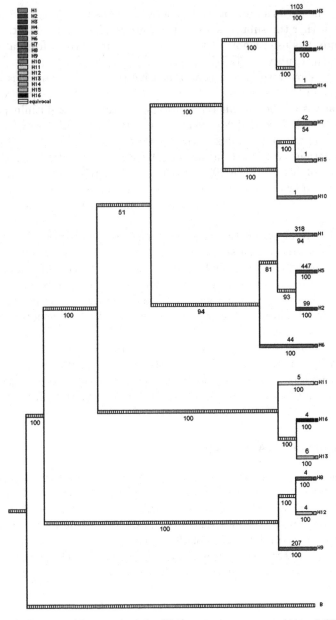

Fig. 2.6. Phylogeny of hemagglutinin (HA) sequences representing 2,358 isolates of influenza A, with a single sequence of influenza B as outgroup. To summarize the source tree we have condensed each subtype clade into a single branch. The numbers of isolates included in the full tree are presented as the numerals above each branch. The numerals below each branch represent jackknife support values (0 worst to 100 best). Sequence and character data was drawn from Genbank (www.ncbi.nlm.nih.gov) and the Influenza Sequence Database (www.flu.lanl.gov)

and mammalian strains of influenza in the trachea of pigs, it has been hypothesized that domestic swine act as intermediate hosts in which human and avian viruses can recombine [SCH90]. This mechanistic hypothesis of viral transmission is widespread. However, as discussed above, a number of events of suspected direct transmission of avian influenza viruses to humans have been reported [LIP04, UNG05].

Hypotheses on the relative frequency of host shifts can be made on a phylogenetic tree through the optimization [FIT71] of a character with states representing various hosts of the viral isolates under consideration (see Sect. 2.2.1). We performed this analysis on our tree of 2,359 HA sequences and found that most of the internal nodes close to the root are optimized as having an avian origin (Fig. 2.7). Thus the results of this analysis are consistent with the hypothesis of an avian origin of all influenza type A viruses [WEB92]. These results also show that most major lineages of influenza A that infect domestic birds originated in aquatic birds. This is compatible with the hypothesis that wild aquatic birds as the natural reservoir of influenza viruses of type A [WEB92].

However, the pattern of host shifts resulting from our study of 2,359 HA sequences seems to be much more complex than previously thought [GAM90, LIP04]. For instance, in many cases, after the spread of influenza type A viruses into domestic bird and mammal populations (including humans), some derived lineages are later spread again to aquatic birds. Furthermore, the results indicate that direct shifts from avian to human hosts have occurred 18–27 times independently in different lineages (without observed intermediate hosts). It must be noted that the possibility of an intermediate host in avian-to-human transmission events cannot be completely rejected. It is possible that an intermediate host existed in nature but it was not sampled by the surveillance program and therefore not included in the analysis. However, based on the available evidence, it seems that host shifts from birds to humans have been frequent in the evolutionary history of influenza type A. Moreover, avian-to-human shifts are more common than swine to human shifts in the history of influenza.

Multiple direct avian-to-human shifts appear to occur in the case of the putative pandemic strains of influenza A (subtype H5N1) that have spread across Eurasia since 1997 [WHO05]. In addition to being highly pathogenic, these H5N1 strains have independently infected other hosts such as felids and pigs in several instances.

Predictive Power of Phylogenetics Analysis

Phylogenetics is practiced by most as a historical science; however, several researchers noted that aspects of the tree shape may be used in predicting future genetic lineages of influenza against which it is important to design vaccines [GRE04]. Notable among these assertions are the studies in the shape of influenza A phylogeny as viewed through the hemagglutinin (HA) gene

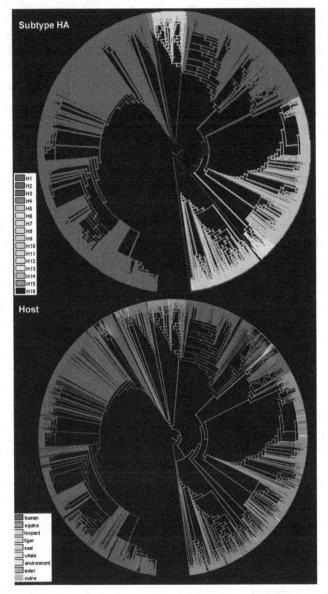

Fig. 2.7. Two character optimizations on the for hemagglutinin (HA) sequences representing 2,358 isolates of influenza A, with an influenza B outgroup at the root. The top tree has an optimization of the character "HA antigenic subtype". The lower tree depicts optimization of the character "host". Character data was drawn from Genbank (www.ncbi.nlm.nih.gov) and the Influenza Sequence Database (www.flu.lanl.gov). Optimizations and tree graphics were made with Mesquite (www.mesquiteproject.org). For better visualization contact the authors for files in scalable pdf format

[BUS99, FER02]. The HA gene codes for a surface glycoprotein of the virion responsible for binding to sialic acid on host cell surface receptors. At a genomic level, lineages of influenza are constantly changing due to mutation that occurs at high rates in RNA viruses. Extinction of evolutionary lineages of viruses to which hosts have become immune or when susceptible hosts are in short supply is common [GRE04]. This process of constant replacement of influenza lineages produces a characteristic coniferous shape to a phylogeny reconstructed from HA sequences[BUS99]. The "conifer" metaphor refers to the hypothesis that influenza HA is constantly changing but there is limited diversity at any time [FER02]. Thus an influenza HA tree appears to be formed by addition of strains to the apex of the tree's trunk that contains the contemporary "infectious" viruses rather than more basal presumably "extinct" lineages to which hosts are immune. Other groups of researchers have used the assumption that there is limited influenza A diversity at any one time to downplay the utility of phylogenetic approaches [PLO02]. As an alternative to phylogenetics, which they consider difficult, these groups make predictions based on size of various clusters of related isolates, termed "swarms" [PLO02]. Several groups, whether using trees or swarms, have identified putatively dominant strains of influenza to predict the genetic makeup of future viral populations [PLO02] [BUS99].

If these assumptions were never violated, the diversity of a previous year's flu season could be assessed, forthcoming strains predicted, and thus used to inform vaccine design. In practice, the CDC uses a mixture of viral strains comprised of H1N1 and H3N2 of influenza A and an influenza B virus. For example, the 2005–2006 vaccine was based on A/New Caledonia/20/99 (H1N1), A/California/7/2004 (H3N2), and B/Shanghai/361/2002 viruses [PAL06]. Notably the H5N1 strain (or any of the other avian strains with potential to infect humans) is currently not considered in the vaccine that is seasonally administered to civilians in the United States.

The ability to predict influenza viral strains that will affect human and animal populations is important. However, prediction methods and experimental designs that are relevant to those methods are in their infancy. Current surveillance programs are focused on detection of antigenically novel strains. As such, surveillance programs are not designed as ecological experiments to quantitatively measure strain-specific incidence and cluster size. Furthermore, the current sample of influenza diversity may be biased by partial genomic sequencing, differences in effort within various geographic and political boundaries, focus on certain subtypes of interest, and differential efforts over time due to variable public concern. Recent papers using whole genome data have indicated that the conifer like growth assumption of HA-based phylogenies that has been central to predictive models of H3N2 seasonal influenza [BUS99, FER02] may be violated. Full genome analysis of H3N2 has shown that there are multiple co-circulating lineages; some of which may be overlooked by vaccine designs [HOL05, GHE05]. Similarly, our large scale-analysis of 2,359 HA sequences depicts that many subtypes and lineages within subtypes of influenza are circulating and being exchanged among human and animal populations at any one time Fig. 2.7.

Viral Surveillance Quality: A Phylogenetic Perspective

In addition to providing hypotheses on the relationships of a group of organisms, phylogenetic trees imply a temporal order of the successive internal nodes (i.e., the time at which a single evolutionary lineage splits producing two independent descendent lineages). Minimal estimates on the date at which these evolutionary splits occur can be obtained through the analysis of the time at which the descendant organisms (leaves) are known to occur. These estimates can be computed with the implementation of an irreversible Sankoff character in which the cost of transformation between two character states represents the amount of time elapsed between the time of appearance of two terminal taxa [PN01].

Influenza A viral sequences are named with the host, locale, and year in which each isolate was sampled by the surveillance program. Several methods exist to measure the correlation between the temporal dates of sampled organisms and the relative order they show in the phylogenetic tree. Here we adapt the Manhattan Stratigraphic Metric (MSM*) to influenza surveillance. The MSM* was originally developed to assess the quality of the fossil record [PN01] (Table 2.2). However, the MSM* is simply a quantitative measure of how well the available data reflects the diversification pattern of the taxa present in the optimal phylogenetic trees and is thus of general utility.

An extensive sampling of sequences, such as the one gathered for the study of 2,359 isolates, is critical to comparatively assess quality of surveillance in various regions, among various strains, and over periods of time. Our results show that this correlation between branching pattern and dates of viral isolation is good in that it significantly differs from a random expectation. This is true over the entire tree as well as when some individual lineages are measured. However, the relative quality of surveillance differs markedly between lineages.

Table 2.2. Results for Manhattan Stratigraphic Metric (MSM*) applied to various subtype clades isolated in the past 30 years

Strain	MSM*
H5N2	0.12
H9N2	0.36
H1N1	0.41
H3N2	0.49
H5N1	0.53

A score close to 1 in the MSM* reflects good surveillance and values close to 0 imply poor surveillance.

One example of differential surveillance quality occurs in two closely related groups of avian influenza of H5 hemagglutinin subtype. One group in this example contains the highly pathogenic H5N1 strains that currently circulate in Eurasia, the Middle East, and Africa. This large clade has been the focus of intense surveillance since the discovery of widespread infection among wild and domestic birds and some avian-to-human transmission[YUA02, UNG05]. The H5N1 viral isolates form a sister clade to H5N2 known from domestic and wild bird in the Americas (H5N2). The number of available hemagglutinin sequences of H5N2 comprise less than one fifth of the number of HA sequences for H5N1. This in itself represents a measure of the surveillance intensity devoted to these two groups of avian influenza. However, even if the number of sequences is normalized at 100 sequences to perform the MSM test, the surveillance quality of the H5N1 clade is far superior to the H5N2 clade.

We can also use visualization techniques to assess surveillance quality. Typically branches of a phylogenetic tree are scaled used to depict the number of mutations or other character changes assigned to each branch. However, we have adapted this use of branch scaling to reflect the number of years that have passed between sampling of related isolates rather than mutations or characters. Compare Fig. 2.8 which has short branch lengths reflecting good surveillance quality with Fig. 2.9 which has long branch lengths implying poor surveillance quality.

Cases in which there is poor correlation between the date of sampling of a given isolate and its inferred date of origin would indicate that the surveillance program is failing to closely monitor the persistence of diverse lineages of influenza (2.9).

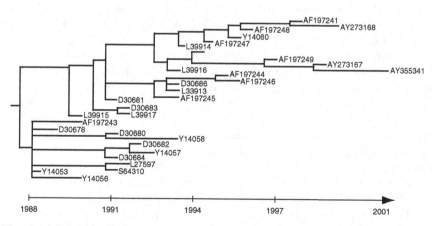

Fig. 2.8. Cases of efficient surveillance in which the sampling of sequences through time reflects the diversification patterns of the phylogenetic tree based on the hemagglutinin sequences. For a full tree contact the authors for files in scalable formats such as pdf and nexus

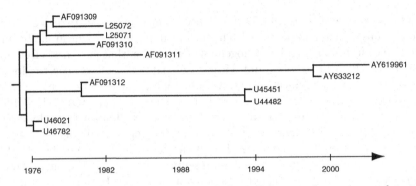

Fig. 2.9. Cases of poor surveillance in which viral lineages have gone undetected for many years since its evolutionary origins. In this case, sequences AY619961 and AY633212 were detected around the year 2000 but their evolutionary origin dates back to the late 1970s. The available sampling of isolates through time does not reflect the diversification patterns of the phylogenetic tree. For a full tree contact the authors for files in scalable formats such as pdf and nexus

2.4 Evolution of SARS Coronaviruses

The Zoontic Origins of Coronaviruses Associated with SARS

Severe Acute Respiratory Syndrome (SARS) is a novel human illness caused by a previously unrecognized coronavirus (CoV) termed SARS-CoV [MAR03] [ROT03]. SARS-CoV may be of zoonotic origin. However as of today, there remain conflicting reports on the zoonotic origins of SARS CoV. Guan et al., 2003 [GUA03] report that SARS CoV originated in small carnivores whereas Li et al., 2005 [LI05] and Lau et al., (2005) [LAU05] counter that SARS CoV originated in bats.

No matter the type of phylogenetic perspective they may espouse, most virologists produce the same basic data by surveying putative host animals and patients with antibodies, then isolating and sequencing partial or whole genomes of various viruses detected in hosts. Molecular phylogenetic analyses of the nucleotide or inferred amino acid sequence data from various viral isolates can then be used to reconstruct the history of the transmission events the virus among hosts. The fundamental belief associated with this research program is that the branching pattern of the phylogeny will reveal a temporal series of transformations when character of interest such as the host is optimized on the viral phylogeny.

Most virology researchers rely on distance methods. The most popular distance method among virologists is neighbor-joining (NJ) [SAI87]. Distance methods require a precomputed multiple alignment of DNA or amino acid sequences drawn from homologous genes of the viral strains of interest. Then in NJ, the most grossly similar pair of isolates (as represented by sequences) are clustered. The clustered pair is then considered as a single taxon and the next most similar pair of taxa is to cluster until only two taxa remain and are joined. In distance methods no outgroups are proposed and no assumptions of ancestral character states are considered. As a result, polarity of transformations can only be inferred as from dissimilar to similar. Distance methods output a single unrooted, star-shaped graph.

Nevertheless preparing figures, some investigators who use distance methods choose to impart directionality by selecting an edge of the graph to serve as a root of the tree. The choice of root is crucial in depicting the polarity of host shifts and depicting clades. The rooting step has been executed variably by researchers comparing sequence data from CoVs isolated from humans with sequence data CoVs isolated from small carnivores. In the case of Guan et al. ([GUA03] see their Figs. 2 and S2 of their supplemental materials) and the Chinese SARS Molecular Epidemiology Consortium ([CSARS04] their supplemental Fig. S7) these researchers simply force the root position on their drawings such that they represent SARS-CoV isolates from small carnivore hosts as ancestral. In other drawings, no outgroup is designated ([CSARS04] their Fig. 2) or a human SARS-CoV outgroup is used and the animal SARS-CoV isolates are omitted from the tree ([CSARS04] their Fig. S6 of the supplemental materials). In the case of Song et al., trees are presented as unrooted ([SON05] their Fig. 1a), rooted on a clade comprised of two SARS-CoV isolates, one from human and the other from a carnivor ([SON05] their Fig. 1b), and in a regression analysis a date for a common ancestor of SARS-CoV isolated from humans is calculated using a human basal group ([SON05] their Fig. 3).

In papers comparing sequence data from SARS-Like CoV recently isolated from bats to that from humans and small carnivores Lau et al.,[LAU05] do not root their trees (their Fig. 2) and Li et al., [LI05] and force the root position on their drawing such that one of the bat sequences is ancestral (their Fig. S4 of the supplemental material). Thus, although all these studies employ distance methods, the researchers use various, often facultative means to infer the animal origins of SARS-CoV.

2.4.1 Importance of Outgroups

Other methods of phylogenetic analysis focus on characters, states, edit costs for changes among states, outgroup assumptions, and polarity of change among states – rather than gross similarity in the case of distance methods. Characters can be polymorphisms recognized in columns of aligned nucleotides or amino acids from sequences of interest or phenotypic states such as host, date of isolation, or antigenic subtype. Another feature of most character based methods that differs from NJ is that character based methods examine many randomly generated trees (each representing an evolutionary hypothesis of character transformations and organismal relationships). Thus the concepts of optimality and hypothesis testing are tightly associated with cladistic and maximum likelihood inference. Optimal trees represent more defensible hypotheses. Moreover, character based methods of molecular phylogenetics rely on explicit choices of outgroup to make assumptions about ancestral character states and thus polarize transformations of phenotypes and genotypes that can be reconstructed from data. In order to make an explicit assumption of ancestral character states the investigator designates at least one taxon as the outgroup. The outgroup method originated in cladistics [WAT81] and has become central to the phylogenetic inference [NIX94]. If chosen carefully, the outgroup estimates baseline character states in sequence and phenotypes such that transformations (such as host shifts) can be reliably inferred. To illustrate the choice of outgroup taxon or taxa and clarify the relationships of the organisms, character based trees are often rooted from the outgroup and ingroup.

Just as in distance methods, in most character-based methods, sequence data is aligned before the phylogenetic analysis. Novel implementations, termed direct optimization, allow unaligned sequence data to be analyzed without precomputing an alignment, Wheeler [WHE96]. In direct optimization, sequence data are aligned as various trees are built and their optimality is assessed (using maximum likelihood and cladistic optimality criteria as specified by the investigator). Thus for each tree a specific alignment that is optimal for that tree is constructed. One additional advantage of direct optimization is that the outgroup need not be designated by the investigator but rather randomized during the search for optimal trees and alignments. In some implementations of character based methods where prealignment is necessary, the outgroup can be randomized by scripting a series of analyses. Outgroup randomization enables analyses of taxa where previous knowledge of ingroup/outgroup relationships is lacking or is among the hypotheses the investigator wants to test via tree search.

2.5 Discussion

Large-scale phylogenetic analyses are particularly useful to study global problems of infectious disease. However, phylogenetic analysis of large number of organisms and whole genomes is an extremely challenging computational problem. Recent advances in heuristic tree search algorithms, alignment methods, and parallel computing strategies have been successful. These advances have pushed upward the limits of taxon sampling considered tractable.

Large data sets analysis is interesting not only because it presents interesting computational challenges, moreover large dataset analysis is leading to new knowledge about natural phenomena. For example, in the recent past, researchers working on small datasets argued that influenza had limited diversity at any one time and that this should allow us to predict which strains are important for vaccine design. On the contrary, with large datasets, we find that there are multiple co-circulating lineages at any one time. Thus, large datasets and means to analyze them are important for future vaccine design.

Character-based approaches to phylogenetics provide a wide variety of tools that can be used to better understand the evolutionary processes underlying the spread of infectious diseases. We have discussed here only some of the wide array of applications that phylogenetics and outgroup criteria can have on genomic studies of infectious disease. We look forward to the continued synergistic development of technological and scientific means to better understand infectious and zoonotic disease.

Acknowledgements

Mr. Farhat Habib M.S. was instrumental to build and maintain cluster computers. Rebecca Allen and Jiarui Lian helped to organize phenotype and genetic data. The authors have no competing interests, financial or otherwise. DP acknowledges the National Science Foundations (NSF) support of the Mathematical Biosciences Institute (MBI) and the MBI. DP and DJ acknowledge the support of the Department of Biomedical Informatics and the Ohio State University Medical Center. DJ acknowledges that this material is based upon work supported by, or in part by, the US Army Research Laboratory and the US Army Research Office under contract/grant number W911NF-05-1-0271 and NSF 0531763.

References

[ANT06]. J. Antonovics, M.E. Hood, and C.H. Baker, Molecular virology: was the 1918 flu avian in origin? Arising from: J.K. Taubenberger et al. Nature **437**, 889–893, Nature **440** (2006) E9.

[ARI343]. Aristotle, Historia Animalium, 343 BC.

[BRA02]. M.J. Brauer, M.T. Holder, L.A. Dries, D.J. Zwickl, P.O. Lewis, and D.M. Hillis, Genetic Algorithms and Parallel Processing in Maximum-Likelihood Phylogeny Inference, Mol. Biol. Evol. **19** (2002) 1717–1726.

[BU99]. R.M. Bush, W.M. Fitch, C.A. Bender, and N.J. Cox, Positive Selection on the H3 Hemagglutinin Gene of Human Influenza Virus A., Mol. Biol. Evol **16** (1999) 1457–1465.

[BUD03]. B. Budowle, M.W. Allard, M.R. Wilson, and R. Chakraborty, Forensics and mitochondrial DNA: Applications, Debates, and Foundations, Annu. Rev. Genomics Hum. Genet. **4** (2003) 119–141.

[BUS04]. R.M. Bush, Influenza as a model system for studying the crossspecies transfer and evolution of the SARS coronavirus, Phil. Trans. R. Soc. Lond. B. **359** (2004) 1067–1073.

[BUS99]. R.M. Bush, C.A. Bender, K. Subbarao, N.J. Cox, and W.M. Fitch, Predicting the Evolution of Human Influenza A., Science **286** (1999) 1921–1925.

[CAR94]. J.P. Carulli, D.M. Chen, W.S. Stark, and D.L. Hartl, D. Phylogeny and physiology of Drosophila opsins., J. Mol. Evol. **38** (1994) 25–62.

[CER98]. C. Ceron, J. Dopazo, E. Zapata, J. Carazo, and O. Trelles, Parallel implementation of DNAml program on message-passing architectures, Parallel Computing **24** (1998) 701–716.

[CHA00]. B. Chang, and M. Donoghue, Recreating ancestral proteins, Trends Ecol. Evol. **15** (2004) 109–114.

[CHA01]. M.A Charleston, Hitch-Hiking: A Parallel Heuristic Search Strategy, Applied to the Phylogeny Problem, J. Comput. Biol. **8** (2001) 79–91.

[CSARS04]. The Chinese SARS Molecular Epidemiology Consortium: Molecular Evolution of the SARS Coronavirus During the Course of the SARS Epidemic in China, Science **303** (2004) 1666–1669.

[DUT05]. G. Dutton, Preparing for a Potential Pandemic. Therapeutic and Vaccine Manufacturers Working to Combat the Avian Flu, Genetic Engineering News **25** (2005) 15:1.

[EAR02]. D. Earn, J. Dushoff, and S. Levin, Ecology and evolution of the flu, T.R.E.E **17** (2002) 334–340.

[FAN02]. T. Fanning, R. Slemons, A. Reid, T. Janczewski, J. Dean, and J. Taubenberger, 1917 Avian influenza Virus Sequences Suggest that the 1918 Pandemic Virus Did Not Acquire Its Hemagglutinin Directly from Birds, Journal of Virology **76** (2002) 7860–7862.

[FAR70]. J.S. Farris, Methods for Computing Wagner trees, Syst. Zool. **19** (1970) 83–92.

[FAR83]. J.S. Farris, The logical basis of phylogenetic analysis. In: Platnick, N.I., Funk, V.A. (eds), Advances in Cladistics, Columbia University Press, New York, 1983.

[FAR96]. J.S. Farris, V.A. Albert, M. Kallersjo, D. Lipscomb, and A.G. Kluge, Parsimony Jackknifing Outperforms Neighbor-Joining, Cladistics **12** (1996) 99–124.

[FEL73]. J. Felsenstein, Maximum Likelihood and Minimum-Step Methods for Estimating Trees from Data on Discrete Characters, Syst. Zool. **22** (1973) 240–249.

[FEL78]. J. Felsenstein, The Number of Evolutionary Trees, Systematic Zoology **27** (1978) 27–33.

[FEL81]. J. Felsenstein, Evolutionary trees from dna sequences: A maximum likelihood approach, J. Mol. Evol. **17** (1981) 368–376.

[FER02]. N.M. Ferguson and R. Anderson, Predicting evolutionary change in the influenza A virus, Nat Med. **8** (2002) 562–3.

[FER03]. N.M. Ferguson, A.P. Galvani, and R.M. Bush, Ecological and immunological determinants of influenza evolution, Nature **422** (2003) 428–433.

[FG82]. L.R. Foulds and R.L. Graham, The Steiner problem in phylogeny is NP-complete, Adv. Appl. Math. **3** (1982) 43–49.

[FIT71]. W.M. Fitch, Towards defining the course of evolution: Minimum change for a specific tree topology, Syst. Zool. **20** (1971) 406–416.

[FIT97]. W.M. Fitch, R.M. Bush, C.A. Bender, and N.J. Cox, Long term trends in the evolution of H(3) HA1 human influenza type A, Proc. Natl. Acad. Sci. USA **94** (1997) 7712–7718.

[FIT83]. W.M. Fitch and T. Smith, Optimal sequence alignments, Proc. Natl. Acad. Sci. USA **80** (1983) 1382–1386.

[FLE05]. R. Fleissner, D. Metzler, and A.V. Haeseler, Simultaneous Statistical Alignment and Phylogeny Reconstruction, Syst. Biol. **54** (2005) 548–561.

[FRA02]. D. Franz, The potential bioweaponization of zoonotic diseases. pp. 15–17 in The Emergence of Zoonotic Diseases: Understanding the Impact on Animal and Human Health. T. Burroughs, S. Knobler, and J. Lederberg, eds., Forum on Emerging Infections Board on Global Health (BGH), Institute of Medicine (IOM). National Academy of Sciences Press, Washington D.C., USA, 2002.

[GAM90]. M. Gammelin, A. Altmller, U. Reinhardt, J. Mandler, V.R. Harley, P.J. Hudson, W.M. Fitch, and C. Scholtissek, Phylogenetic analysis of nucleoproteins suggests that human influenza A viruses emerged from a 19th century avian ancestor, Mol. Biol. Evol. **7** (1990) 194–200.

[GER05]. J. Gerberding, Pandemic Planning and Preparedness, http://www.cdc.gov/Washington/testimony/in05262005.htm (2005)

[GHE05]. E. Ghedin, N. Sengamalay, M. Shumway, J. Zaborsky, T. Feldblyum and 14 others, Large-scale sequencing of human influenza reveals the dynamic nature of viral genome evolution, Nature **437** (2005) 1162–1166.

[GIB06]. M. Gibbs and A. Gibbs, Was the 1918 pandemic caused by a bird flu? Arising from: J.K. Taubenberger et al. Nature **437**, 889–893, Nature (2005) 440:E8.

[GOL02]. P.A. Goloboff, W.C. Wheeler, and D. Pol, Parallel searches of large datasets, Cladistics **19** (2002) 151.

[GOL03]. P.A. Goloboff, S.J. Farris, and K.C. Nixon, TNT: Tree Analysis Using New Technologies, Software package distributed by the authors and available at: http://www.zmuc.dk/public/phylogeny/TNT (2003)

[GOL99]. P.A. Goloboff, Analyzing Large Datasets in Reasonable Times: Solutions for Composite Optima, Cladistics **15** (1999) 415–428.

[GRA03]. T. Grant and A. Kluge, Data exploration in phylogenetic inference: Scientific, heuristic, or neither, Cladistics **19** (2003) 379–418.

[GRE04]. B. Grenfell, O. Pybus, J. Gog, J. Wood, J. Daly, J. Mumford, and E. Holmes, Unifying the epidemiological and evolutionary dynamics of pathogens, Science **303** (2004) 327–332.

[GUA03]. Y. Guan, B. Zheng, Y. He, X.L. Liu, Z.X. Zhuang, and 13 others, Isolation and characterization of viruses related to the SARS coronavirus from animals in southern China, Science **302** (2003) 276–278.

[HEN66]. W. Hennig, Phylogenetic Systematics, University of Illinois Press, Urbana, 1966.

[HEN82]. M.D. Hendy and D. Penny, Branch and bound algorithms to determine minimal evolutionary trees, Math. Biosc. **59** (1982) 277–290.

[HHS04]. National Vaccine Program Office, Pandemics and Pandemic Scares in the 20th Century, http://www.hhs.gov/nvpo/pandemics/flu3.htm (2004)

[HHSb]. HHS Pandemic Influenza Plan, Part 2 Public Health Guidance for State and Local Partners, http://www.hhs.gov/pandemicflu/plan/pdf/S02.pdf (Date)

[HIL03]. D.M. Hillis, D.D. Pollock, J.A. McGuire, and D.J. Zwickl, Is Sparse Taxon Sampling a Problem for Phylogenetic Inference?, Syst. Biol. **52** (2003) 124–126.

[HIL96]. D.M. Hillis, Inferring Complex Phylogenies, Nature **369** (1996) 130–131.

[HOL05]. E. Holmes, E. Ghedin, N. Miller, J. Taylor, Y. Bao, and 6 others, Whole-Genome Analysis of Human Influenza A Virus Reveals Multiple Persistent Lineages and Reassortment among Recent H3N2 Viruses, PLoS Biol. **3** (2005) 1–11.

[HUE02]. J. Huelsenbeck, B. Larget, R. Miller, and F. Ronquist, Potential applications and pitfalls of Bayesian inference of phylogeny, Syst. Biol. **51** (2002) 673–688.

[JAN01]. D. Janies and W.C. Wheeler, Efficiency of parallel direct optimization. Cladistics, **17** (2001) S71–S82.

[JAN02]. D. Janies and W.C. Wheeler, Theory and practice of parallel direct optimization. pp. 115–124 in R. Desalle, G. Giribet and W. Wheeler eds. Molecular Systematics and Evolution: Theory and Practice, Birkhuser Verlag, Basel Switzerland, 2002.

[JON03]. R. Johnston, Integrating Methodologists into Teams of Substantive Experts, Studies in Intelligence **47** (2003) No. 1.

[JON95]. J.A. Jones, K.A. Yelick, Parallelizing the Phylogeny Problem, Proc. 1995 ACM/IEEE Conf. Supercomp., **25** (1995)

[KOO04]. M. Koopmans, B. Wilbrink, M. Conyn, G. Natrop, H. van der Nat, H. Vennema, A. Meijer, J. van Steenbergen, R. Fouchier, A. Osterhaus, and A. Bosman, Transmission of H7N7 avian influenza A virus to human beings during a large outbreak in commercial poultry farms in the Netherlands, The Lancet **363** (2004) 587–593s.

[KSI03]. T. Ksiazek, D. Erdman, C. Goldsmith, S. Zaki, T. Peret and 22 others, A Novel Coronavirus Associated with Severe Acute Respiratory Syndrome, The New England Journal of Medicine **348** (2003) 1953–1966.

[LAU05]. S. Lau, P. Woo, K. Li, Y. Huang, H. Tsoi, and 5 others, Severe acute respiratory syndrome coronavirus-like virus in Chinese horseshoe bats, Proc. Natl. Acad. Sci. USA **102** (2005) 14040–14045.

[LAV01]. G. Laver and E. Garman, The Origin and Control of Pandemic Influenza, Science **293** (2001) 1776–1777.

[LEM02]. A.R. Lemmon and M.C. Milinkovitch, The metapopulation genetic algorithm: An efficient solution for the problem of large phylogeny estimation, Proc. Natl. Acad. Sci. USA **99** (2002) 10516–10521.

[LEW98]. P.O. Lewis, A Genetic Algorithm for Maximum-Likelihood Phylogeny Inference Using Nucleotide Sequence Data, Mol. Biol. Evol. **15** (1998) 277–283.

[LIN00]. Y. Lin, M. Shaw, Y. Gregory, K. Cameron, W. Lim, and 8 others, Avian-to-human transmission of H9N2 subtype influenza A viruses: Relationship between H9N2 and H5N1 human isolates, Proc. Natl. Acad. Sci. USA **97** (2000) 9654–9658.

[LI00]. S. Li, D.K. Pearl, and H. Doss, Phylogenetic Tree Construction using Markov Chain Monte Carlo, J. Am. Stat. Assoc. **95** (2000) 493–508.

[LI03].   K. Li, ClustalW-MPI: a parallel implementation of Clustal-W, based on MPI Bioinformatics **19** (2003) 1585–1586.

[LI04].   K. Li, Y. Guan, J. Wang, G. Smith, K. Xu, and 17 others, Genesis of a highly pathogenic and potentially pandemic H5N1 influenza virus in eastern Asia, Nature **430** (2004) 209–213.

[LI05].   W. Li, Z. Shi, M. Yu, W. Ren, C. Smith, and 12 others, Bats are natural reservoirs of SARS-like coronaviruses, Science **310** (2005) 676–679.

[LIP04]. Lipatov et al, Influenza Emergence and Control, J. Virol. (2004) 8951–8959.

[MAR03]. M.A. Marra, S.J. Jones, C.R. Astell, R.A Holt RA, and 47 others, The Genome Sequence of the SARS-Associated Coronavirus, Science **300** (2003) 1399–404.

[MAR04]. B.E. Martina, B.L. Haagmans, T. Kuiken, R.A. Fouchier, G,F. Rimmelzwaan and 5 others, SARS infection of cats and ferrets, Nature **425** (2003) 915.

[MET02]. M. Metzker, D. Mindell, X. Liu, R. Ptak, R. Gibbs, and D. Hillis, Molecular evidence of HIV-1 transmission in a criminal case, Proc. Natl. Acad. Sci. USA **99** (2002) 14292–14297.

[MOI99]. A. Moilanen, Searching for most parsimonious trees with simulated evolutionary optimization, Cladistics **15** (1999) 39–50.

[MOR97]. D. Morrison and J. Ellis, Some effects of nucleotide sequence alignment on phylogeny estimation, Molecular Biology and Evolution **14** (1997) 428–441.

[MOR95]. S. Morse, Factors in the Emergence of Infectious Diseases, Emerging Infectious Diseases **1** (1995) 7–15.

[NIX94]. K.C. Nixon and J.M. Carpenter, On outgroups, Cladistics **9** (1994) 413–426.

[NIX99]. K.C. Nixon, The Parsimony Ratchet, a New Method for Rapid Parsimony Analysis, Cladistics **15** (1999) 407–414.

[OBE06]. E. Obenauer, J. Denson, P. Mehta, X. Su, S. Mukatira, and 12 others, Large-Scale Sequence Analysis of Avian Influenza Isolates, Science, published online January 26 http://www.sciencemag.org/cgi/content/full/1121586/DC1 (2006)

[PAL06]. P. Palese, Making Better Influenza Virus Vaccines?, Emerging Infectious Diseases **12** (2006) 61–65.

[PHI00]. A. Phillips, D. Janies, and W.C. Wheeler, Multiple sequence alignment in phylogenetic analysis, Mol. Phylogenet. Evol. **16** (2000) 317–330.

[PLO02]. J. Plotkin, J. Dushoff, and S. Levin, Hemagglutinin sequence clusters and the antigenic evolution of influenza A virus, Proc. Natl. Acad. Sci. USA **99** (2002) 6263–68.

[PN01].   D. Pol and M.A. Norell, Comments on the Manhattan Stratigraphic Measure, Cladistics **17** (2001) 285–289.

[POE98]. S. Poe, The Effect of Taxonomic Sampling on Accuracy of Phylogeny Estimation, Test Case of a Known Phylogeny, Mol. Biol. Evol. **15** (1998)1086–1090.

[RAN96]. B. Rannala and Z. Yang, Probability distribution of molecular evolutionary trees: a new method of phylogenetic inference, J. Mol. Evol. **43** (1996) 304–311.

[RAN98]. B. Rannala, J.P. Huelsenbeck, Z. Yang, and R. Nielsen, Taxon Sampling and the Accuracy of Large Phylogenies, Syst. Biol. **47** (1998) 702–710.

[RED05]. B.D. Redelings and M.A. Suchard, Joint Bayesian estimation of alignment and phylogeny, Syst. Biol. **54** (2005) 401–418.

[RIC97]. K. Rice and T. Warnow, Parsimony is hard to beat, Computing and combinatorics, (Shanghai, 1997): 124–133 (1997)

[ROS02]. R.S. Ross, S. Viazov, and M. Roggendorf, Phylogenetic analysis indicates transmission of hepatitis C virus from an infected orthopedic surgeon to a patient, J. Med. Virol. **66** (2002) 4617.

[ROS04]. U. Roshan, T. Warnow, B.M.E. Moret, and T.L. Williams, Rec-IDCM3: A Fast Algorithmic Technique for Reconstructing Large Phylogenetic Trees, Proc. IEEE Comp. Syst. Bioinf. Conf. (2004)

[ROT03]. P.A. Rota, M.S Oberste, S.S. Monroe, W.A. Nix, R. Campagnoli, and 30 others, Characterization of a novel coronavirus associated with severe acute respiratory syndrome, Science **300** (2003) 1394–9.

[RUB03]. E.M. Rubin and A. Tall, Perspectives for vascular genomics, Nature **407** (2004) 265–269.

[SAI87]. N. Saitou and M. Nei, The neighbor-joining method: a new method for reconstructing phylogenetic trees, Mol. Biol. Evol. **4** (1987) 406–425.

[SAL01]. L.A. Salter and D.K. Pearl, Stochastic Search Strategy for Estimation of Maximum Likelihood Phylogenetic Trees, Syst. Biol. **50** (2001) 7–17.

[SAN83]. D. Sankoff and R. Cedergren, Simultaneous comparison of three or more sequences related by a tree in D. Sankoff and J. B. Kruskal eds. Time Warps, String Edits, and Macromolecules: the Theory and Practise of Sequence Comparison. Addison-Wesley, Reading, MA. (1983) 253–264.

[SCH90]. C. Scholtissek, Pigs as the mixing vessel for the creation of new pandemic influenza A viruses, Med. Principles Practice **2** (1990) 65–71.

[SEA03]. D. Searls, Pharmacophylogenomics: Genes, evolution and drug targets, Nature Reviews Drug Discovery **2** (2003) 613–623.

[SEA96]. D.L Swofford, G.J. Olsen, P.J. Waddell, and D.M. Hillis, Phylogenetic inference. In: Hillis, D.M., Moritz, C. Mable, B.K. (eds) Molecular Systematics, second edition. Sinauer Associates, Sunderland, 1996.

[SIL97]. J. Silvertown, M. Franco, and J.L. Harper, Plant Life Histories Ecology, Phylogeny and Evolution, Cambridge University Press, Cambridge, 1997.

[SNE00]. Q. Snell, M. Whiting, M. Clement, and D. McLaughlin, Parallel Phylogenetic Inference, Proc. 2000 ACM/IEEE Conf. Supercomp., **35** (2000)

[SNE73]. P.H.A Sneath and R.R. Sokal, Numerical taxonomy The principles and practice of numerical classification, W. H. Freeman, San Francisco. xv + 573 p. (1973)

[SON05]. H. Song, C. Tu, G. Zhang, S. Wang, K. Zheng, and 21 others, Crosshost evolution of severe acute respiratory syndrome coronavirus in palm civet and human, Proc. Natl. Acad. Sci. USA **102** (2005) 2430–2435.

[STA02]. A. Stamatakis, T. Ludwig, H. Meier, and M.J. Wolf, Accelerating Parallel Maximum Likelihood-based Phylogenetic Tree Calculations using Subtree Equality Vectors, Proc. 15 IEEE/ACM Supercomp. Conf.(2002)

[STE00]. T. Sterling, J. Salmon, D. Becker, and D. Savarese, How to Build a Beowulf. A Guide to the Implementation and Application of PC Clusters, MIT press, 2000.

[SUZ02]. Y. Suzuki and M. Nei, Origin and Evolution of Influenza Virus Hemagglutinin Genes, Mol. Biol. Evol. **19** (2003) 501–509.

[SWO02]. D.L. Swofford, Paup: Phylogenetic Analysis using Parsimony (and other methods), Sinauer Associates, Sunderland, 2002.

[SWO91]. D.L. Swofford, When are phylogeny estimates from molecular and morphological data incongruent? In: Miyamoto, M.M., Cracraft, J. (eds) Phylogenetic analysis of DNA sequences, Oxford Univ. Press, Oxford, 1991.

[TAU01]. J.K. Taubenberger, A.H. Ried, T.A. Janczeqski, and T.G. Fanning, Integrating historical, clinical and molecular genetic data in order to explain the origin and virulence of the 1918 Spanish influenza virus, Philos. Trans. R. Soc. Lond. B. **356** (2001) 1829–1839.

[TAU97]. J.K. Taubenberger, A. Reid, A.E. Frafft, K.E. Bijwaard, and T. Fanning, Initial Genetic Characterization of the 1918 Spanish Influenza Virus, Science **275** (1997) 1793–1796.

[TAU05]. J.K. Taubenberger, A. Reid, R. Lourens, R. Wang, G. Jin, and T. Fanning, Characterization of the 1918 influenza virus polymerase genes, Nature **437** (2005) 889–893.

[TAU06]. J.K. Taubenberger and D. Morens, 1918 influenza: the mother of all pandemics, Emerg Infect Dis. **12** (2006) 15–22.

[TAY01]. L.H. Taylor, S. M. Latham, and M. E. Woolhouse, Risk factors for human disease emergence, Philos. Trans. R. Soc. Lond. B. Biol. Sci. **356** (2001) 983–989.

[TEH03]. A. Tehler, D.P. Little, J.S. Farris, The full-length phylogenetic tree from 1551 ribosomal sequences of chitinous fungi, Fungi. Mycol. Res. **107** (2003) 901–916.

[THO94]. J.D. Thompson, D.G. Higgins, and T.J. Gibson, CLUSTAL W: improving the sensitivity of progressive multiple sequence alignments through sequence weighting, position specific gap penalties and weight matrix choice, Nucl. Acids Res. **22** (1994) 4673–4680.

[THR04]. J. Thornton, Resurrecting ancient genes, Experimental analysis of extinct molecules, Nat. Rev. Genet. **5** (2004) 366–375.

[TKF92]. J.L. Thorne, H. Kishino, and J. Felsenstein, Inching toward reality: An improved likelihood model of sequence evolution, J. Mol. Evol. **34** (1992) 3–16.

[TWE04]. S. Tweed, D. Skowronski, S. David, A. Larder, M. Petric, and 10 others, Human illness from avian influenza H7N3, British Columbia.

Emerg Infect Dis. http://www.cdc.gov/ncidod/EID/vol10no12/04-0961.htm (2004)

[UNG05]. K. Ungchusak, P. Auewarakul, S. Dowell, R. Kitphati, P. Wattana, and 10 others, Probable Person-to-Person Transmission of Avian Influenza A (H5N1), N Engl J Med **352** (2005) 333–340.

[UKK85]. E. Ukkonen, Algorithms for approximate string matching, Information and Control Archive **64** (1985) 100–118.

[WAN94]. L. Wang and T. Jiang, On the complexity of multiple sequence alignment, J. Comput. Biol. **4** (1994) 337–348.

[WAN05]. Q. Wang, M. Han, J. Funk, G. Bowman, D. Janies, and L. Saif, Genetic Diversity and Recombination of Porcine Sapoviruses. Journal of Clinical Microbiology **43** (2005) 5963–5972.

[WAT81]. L. Watrous and Q. Wheeler, The outgroup comparison method of character analysis, Syst. Zool. **30** (1981) 1–11.

[WEB92]. R.G. Webster, W.J. Bean, O.T. Gorman, T.M. Chambers, and Y. Kawaoka, Evolution and ecology of influenza A viruses, Microbiol. Rev. **56** (1992) 152–179.

[WHE94]. W. Wheeler and D. Gladstein, MALIGN: A multiple sequence alignment program, J. Hered. **85** (1994) 417–418.

[WHE95]. W. Wheeler, Sequence alignment, parameter sensitivity, and the phylogenetic analysis of molecular data, Systematic Biology **44** (1995) 321–331.

[WHE90]. W.C. Wheeler, Nucleic acid sequence phylogeny and random outgroups, Cladistics **6** (1990) 363–367.

[WHE04]. W.C. Wheeler, D. Janies, and J. DeLaet, DNA sequence alignment and parallel processing, McGraw-Hill Yearbook of Science and Technolog, (2004)

[WHE05]. W.C. Wheeler, A. Varon, D. Gladstein, and J. DeLaet, POY. Phylogeny program for optimization of nucleic acids and other data, American Museum of Natural History, http://research.amnh.org/scicomp/projects/poy.php (2005)

[WHE96]. W.C. Wheeler, Optimization Alignment: the end of multiple sequence alignment in phylogenetics?, Cladistics **12** (1996) 1–9.

[WHI03]. K.P. White, Functional genomics and the study of development, variation and evolution, Nature Genetics **2** (2003) 528–537.

[WHO05]. Report on Global Surveillance of Epidemic-prone Infectious Diseases - Influenza (2000)

[WHO07a]. H5N1 avian influenza: timeline of major events; 11 September (2007)

[WHO07b]. WHO. Cumulative Number of Confirmed Human Cases of Avian Influenza A/(H5N1) Reported to WHO; 10 September (2007)

[YAN07]. Y. Yang, M.E. Halloran, J. Sugimoto, Jr I.M Longini, Detecting human-to-human transmission of avian influenza A (H5N1), Emerg Infect Dis. http://www.cdc.gov/EID/content/13/9/1348.htm (2007).

[YUA02]. Y. Guan, J.S.M. Peiris, A.S. Lipatov, T.M. Ellis, K.C. Dyrting, S. Krauss, L.J. Zhang, R.G. Webster, and K.F. Shortridge, Emergence of multiple genotypes of H5N1 avian influenza viruses in Hong Kong SAR, Proc. Natl. Acad. Sci. USA **99** (2002) 8950–8955.

[ZWI02]. D.J. Zwickl and D.M. Hillis, Increased Taxon Sampling Greatly Reduces Phylogenetic Error, Syst. Biol. **51** (2002) 588–598.

Reaction–Diffusion Equations and Ecological Modeling

C. Cosner

Department of Mathematics, University of Miami,
Coral Gables, FL 33124, USA
email: gcc@math.miami.edu

Summary. Reaction-diffusion equations are widely used as models for spatial effects in ecology. They support three important types of ecological phenomena: the existence of a minimal patch size necessary to sustain a population, the propagation of wavefronts corresponding to biological invasions, and the formation of spatial patterns in the distributions of populations in homogeneous environments. Reaction-diffusion equations can be analyzed by means of methods from the theory of partial differential equations and dynamical systems. we will discuss the derivation of reaction-diffusion models in ecology, sketch the basic aspects of their analysis, and describe some of their applications and mathematical properties.

3.1 Introduction

3.1.1 General Description of Models

Reaction–diffusion equations arise as models for the densities of substances or organisms that disperse through space by Brownian motion, random walks, hydrodynamic turbulence, or similar mechanisms, and that react to each other and their surroundings in ways that affect their local densities. Reaction–diffusion models are in themselves deterministic, but they can be derived as limits of stochastic processes under suitable scaling. Specifically, they provide a modeling approach that allows us to translate assumptions about stochastic local movement into deterministic descriptions of global densities. Reaction–diffusion models are spatially explicit, describe population densities, and treat space and time as continuous. These features distinguish them from other types of spatial models such as interacting particle systems, integrodifference models, and metapopulation models. There are three major types of ecological phenomena that are supported by reaction-diffusion equations: the existence of a minimal patch size necessary to support a population, the presence of traveling wavefronts corresponding to biological invasions, and the formation of spatial patterns. In what follows I will discuss how reaction–diffusion models are formulated and how they describe those phenomena. In the course of that discussion I will also describe some of their basic mathematical properties.

3.1.2 Reaction

In the context of ecological models, the reaction terms in reaction–diffusion equations and systems are typically the same as those that are used in non-spatial population models based on ordinary differential equations. Thus, for a single population, the reaction terms would be those that might occur in a model for a population density $P(t)$ of the form

$$\frac{dP}{dt} = f(P), \tag{3.1}$$

where $f(P)$ often has the form $f(P) = g(P)P$. Common choices for $f(P)$ are $f(P) = rP$ (linear growth), $f(P) = r(1 - [P/K])P$ (logistic growth), or $f(P) = r(P - a)(1 - [P/K])P$ with $a \in (0, K)$ (growth with Allee effect). For systems, typical reaction terms are those that occur in nonspatial models for competition, mutualism, or predator–prey interactions. Those include Lotka–Volterra models, but also more general models such as predator–prey models with a functional response. If P and Q are population densities, a Lotka–Volterra competition model would have the form

$$\begin{aligned}
\frac{dP}{dt} &= (A - BP - CQ)P, \\
\frac{dQ}{dt} &= (D - EP - FQ)Q
\end{aligned} \tag{3.2}$$

with A, B, C, D, E, F positive constants. A fairly typical predator prey model would have the form

$$\begin{aligned}
\frac{dP}{dt} &= (A - BP)P - Cf(P, Q), \\
\frac{dQ}{dt} &= -DQ + Ef(P, Q),
\end{aligned} \tag{3.3}$$

where $f(P, Q) = g(P, Q)Q$ with $g(P, Q)$ being the functional response. For Lotka–Volterra models $g(P, Q) = FP$; other common forms are

$$\begin{aligned}
g(P, Q) &= \frac{FP}{1 + HP} && \text{(Holling type II)}, \\
g(P, Q) &= \frac{FP^2}{1 + HP^2} && \text{(Holling type III)}, \\
g(P, Q) &= \frac{FP}{1 + HP + JQ} && \text{(Beddington} - \text{DeAngelis)}.
\end{aligned} \tag{3.4}$$

These sorts of formulations can be extended to encompass more species or more trophic levels, or modified to describe mutualism or other sorts of population interactions. In some cases they can be derived from mechanistic assumptions. For further discussion see [22, 34, 50]. Often an important aspect of the analysis of models such as (3.1)–(3.3) is to find their equilibria and determine the stability of those equilibria. In the case of systems the stability analysis often involves the eigenvalues of matrices obtained by linearizing about the equilibria. Equilibria and eigenvalues play a similar role in the analysis of reaction–diffusion models, but the eigenvalues generally are associated with differential operators rather than matrices.

3.1.3 Diffusion and Random Walks

Diffusion is a description of movement that arises as a result of an object or organism making many short movements in random directions. The diffusive description of random motion emerges as a continuum limit of such random walks when the length Δx of each step and the time Δt required for each step go to zero in such a way that the ratio $(\Delta x)^2/\Delta t$ remains constant. To understand how this works it is useful to consider a simple example in one space dimension. Suppose that an organism moves along a line by moving a distance Δx to the left with probability $1/2$ or a distance Δx to the right with probability $1/2$ at each time step Δt. Suppose that $p(x,t)$ is the probability that the organism is at location x at time t. To arrive at that point at that time it must have been either one step to the left at time $t - \Delta t$ and then moved to the right, or one step to the right and have moved to the left. Thus, we have

$$p(x,t) = \frac{1}{2}p(x + \Delta x, t - \Delta t) + \frac{1}{2}p(x - \Delta x, t - \Delta t). \qquad (3.5)$$

If we subtract $p(x, t - \Delta t)$ from both sides and divide by Δt we obtain

$$\frac{p(x,t) - p(x,t - \Delta t)}{\Delta t} = \frac{1}{2\Delta t}[p(x + \Delta x, t - \Delta t) - 2p(x, t - \Delta t)$$
$$+ p(x - \Delta x, t - \Delta t)]. \qquad (3.6)$$

Suppose that we now impose the diffusive scaling

$$(\Delta x)^2/\Delta t = 2D. \qquad (3.7)$$

Then (3.6) becomes

$$\frac{p(x,t) - p(x,t - \Delta t)}{\Delta t} = \frac{D}{(\Delta x)^2}[p(x + \Delta x, t - \Delta t) - 2p(x, t - \Delta t)$$
$$+ p(x - \Delta x, t - \Delta t)]. \qquad (3.8)$$

The expression on the left is a difference quotient in t; the expression on the right is a second difference in x. Taking the limit of (3.8) as Δx, $\Delta t \to 0$ while (3.7) remains in force yields the diffusion equation

$$\frac{\partial p}{\partial t} = D\frac{\partial^2 p}{\partial x^2}. \qquad (3.9)$$

Mathematically this is identical to the heat equation. Note that the scaling (3.7), where $2D$ is the square of the distance Δx moved by the organism in a time unit Δt, produces a coefficient in front of the term $\partial^2 p / \partial x^2$, which is equal to $1/2$ of the square of the distance moved per unit time. This interpretation of the diffusion coefficient D is valid in any number of dimensions.

Equation (3.9) describes the probable location of a single organism. If we solve (3.9) for $p(x, t)$ corresponding to a single organism starting at time $t = 0$ at location $x = z$ we obtain

$$p(x, t) = \frac{1}{\sqrt{4\pi Dt}} e^{-(x-z)^2/4Dt}. \tag{3.10}$$

If we start with a collection of organisms at $t = 0$ with density $u_0(x)$ then the expected density $u(x, t)$ at time t is obtained by averaging (3.10)

$$u(x, t) = \frac{1}{\sqrt{4\pi Dt}} \int_{-\infty}^{\infty} e^{-(x-y)^2/4Dt} u_0(y) \, dy. \tag{3.11}$$

The expression in (3.10) is the fundamental solution for the heat/diffusion equation (3.9), and so if our organisms are just moving and not dying or reproducing we have

$$\frac{\partial u}{\partial t} = D \frac{\partial^2 u}{\partial x^2}. \tag{3.12}$$

This is the diffusion equation for the density $u(x, t)$, which is the diffusion part of typical reaction–diffusion models. For further discussion of the derivation of diffusion equations see [6, 34, 35, 48]. The formula (11) is derived in many introductory texts on partial differential equations, see for example [31, 47].

In higher space dimensions (3.12) becomes

$$\frac{\partial u}{\partial t} = D\nabla^2 u = D\left(\frac{\partial^2 u}{\partial x^2} + \frac{\partial^2 u}{\partial y^2}\right)$$
$$\text{or} \quad D\left(\frac{\partial^2 u}{\partial x^2} + \frac{\partial^2 u}{\partial y^2} + \frac{\partial^2 u}{\partial z^2}\right). \tag{3.13}$$

The operator ∇^2 is known as the Laplacian or Laplace operator and is sometimes denoted as Δ. The coefficient D in (3.13) still represents $1/2$ the mean square distance traveled by an organism in unit time in all space dimensions. That fact can be used to calculate D from mark-recapture experiments, see [2, 36, 48]. For models in n space dimensions the expression corresponding to (3.10) is

$$\frac{1}{(4\pi Dt)^{n/2}} e^{-|x-z|^2/4Dt}, \tag{3.14}$$

where $|x| = \sqrt{x_1^2 + \cdots + x_n^2}$ is Euclidean distance.

3.1.4 Flux, Fick's Law, and Boundary Conditions

A common and useful way to formulate and interpret diffusion equations and related models for transport or movement is by analogy to the transport of substances by physical diffusion and advection. Suppose that a substance flows and diffuses along a tube. The flux $J(x)$ denotes the rate per unit area at which the substance is transported across an interface on a cross-section of the tube at point x.

Let A denote the cross-sectional area of the tube. Assume that the tube is sufficiently narrow that J is approximately constant on cross-sections. If the substance is at density $u(x, t)$ then the rate of change of the amount of substance in the interval $(x, x + \Delta x)$ over the time interval Δt is given by

$$\frac{(A\Delta x)u(x, t + \Delta t) - (A\Delta x)u(x, t)}{\Delta t} = A[J(x, t) - J(x + \Delta x, t)]. \quad (3.15)$$

Dividing (3.15) by $A\Delta x$ and letting $\Delta x, \Delta t \to 0$ we obtain

$$\frac{\partial u}{\partial t} = -\frac{\partial J}{\partial x}. \quad (3.16)$$

If the transport is by diffusion, Fick's law gives the flux as

$$J = -D\frac{\partial u}{\partial x}. \quad (3.17)$$

(Fick's law is an empirical result about physical diffusion via Brownian motion, but it is a reasonable approximation for other types of biological diffusion.) If the transport arises from advection with velocity v to the right then

$$J = vu. \quad (3.18)$$

Combining (3.17) and (3.18) in (3.16) yields

$$\frac{\partial u}{\partial t} = -\frac{\partial}{\partial x}\left(-D\frac{\partial u}{\partial x} + vu\right)$$

$$= D\frac{\partial^2 u}{\partial x^2} - \frac{\partial}{\partial x}(vu). \quad (3.19)$$

The advective terms in (3.19) and (3.20) might correspond to physical advection in aquatic environments, but they could also arise as descriptions of directed movement along environmental gradients. In the case $D = 0$, $v = \text{constant}$, (3.19) becomes $\dfrac{\partial u}{\partial t} = -v\dfrac{\partial u}{\partial x}$, which has solutions $u = g(x - vt)$ which maintain a constant profile but move to the right at velocity v. In higher space dimensions flux and velocity are vectors. The analog of (3.19) is

$$\frac{\partial u}{\partial t} = -\nabla \cdot \mathbf{J} = -\nabla \cdot [-D\nabla u + \mathbf{v}u]$$

$$= \nabla \cdot D\nabla u - \nabla \cdot (\mathbf{v}u). \quad (3.20)$$

If $D = \text{constant}$ and $\mathbf{v} = 0$ this reduces to (3.13).

For models involving diffusion in a region $\Omega \subseteq \mathbb{R}^n$ with boundary $\partial\Omega$, it is necessary to specify what happens at the boundary. Typical boundary conditions involve specifying the density at the boundary, the flux across the boundary, or a relation between those. If the density is specified we have a Dirichlet boundary condition

$$u(x,t) = g(x) \tag{3.21}$$

(here x represents a point $(x_1, \ldots, x_n)$ on the boundary $\partial\Omega$ of Ω.) If the flux is specified we obtain from Fick's law as in (3.20)

$$\mathbf{J} \cdot \mathbf{n} = (-D\nabla u + \mathbf{v}u) \cdot \mathbf{n}$$

$$= -D\frac{\partial u}{\partial \mathbf{n}} + (\mathbf{v} \cdot \mathbf{n})u = g(x) \quad \text{on } \partial\Omega, \tag{3.22}$$

where $\mathbf{n}$ is the outward pointing unit normal vector on $\partial\Omega$, $\partial u/\partial \mathbf{n}$ is the directional derivative of u in the direction of $\mathbf{n}$, and $\mathbf{J}$ is the flux vector. If there is no advection ($\mathbf{v} = 0$) then (3.22) becomes a Neumann condition. A common relation that might be specified between density at $\partial\Omega$ and flux across $\partial\Omega$ is that the flux should be proportional to the density, so that

$$-D\frac{\partial u}{\partial \mathbf{n}} + (\mathbf{v} \cdot \mathbf{n})u = \gamma u, \tag{3.23}$$

with $\gamma > 0$. This type of condition is called a Robin condition.

The no-flux boundary condition (3.22) with $g(x) = 0$ can be interpreted as saying that nothing crosses the boundary of the region Ω. In ecological models this would correspond to a situation where organisms inhabiting a patch do not leave when they reach the boundary. If $g(x) = 0$ in the Dirichlet condition (3.21), the interpretation is that anything that reaches the boundary leaves or is removed and does not return. This is often viewed in ecological models as describing a patch surrounded by a region that is immediately lethal to the organisms inhabiting the patch. The Robin condition (3.23) can be interpreted as saying that when organisms reach the patch boundary some leave the patch but some do not.

3.1.5 Reaction–Diffusion Models

we can now formulate reaction diffusion models. Let Ω be a region in 1, 2, or 3 dimensions, with boundary $\partial\Omega$. Let $x = (x_1, \ldots, x_n)$ denote a point in $\bar{\Omega}$, and let $u(x,t)$ denote the density of a substance or population on Ω at time t. A typical reaction–diffusion model for u would have the form

$$\frac{\partial u}{\partial t} = D\nabla^2 u + f(u) \quad \text{in } \Omega \times (0, \infty)$$

$$\alpha\frac{\partial u}{\partial \mathbf{n}} + (1-\alpha)u = 0 \text{ on } \partial\Omega \times (0, \infty) \tag{3.24}$$

$$u(x,0) = h(x) \quad \text{on } \Omega,$$

where $0 \leq \alpha \leq 1$.

Remark on Notation. It is common for mathematicians to denote the Laplace operator ∇^2 as Δ. The differential equation in (3.24) then has the form

$$\frac{\partial u}{\partial t} = D\Delta u + f(u).$$

If the function $f(u)$ and the boundary $\partial\Omega$ of Ω are smooth and $h(x)$ is continuous on $\bar{\Omega}$ the problem (3.24) will be well posed in the sense that it will have a unique solution $u(x,t)$ on $\Omega \times (0,T]$ for some $T > 0$, with $u(x,t)$ depending continuously on $h(x)$. If $T < \infty$ then we must have $\max_{\bar{\Omega}}|u(x,t)| \to \infty$ as $t \uparrow T$. (These are relatively standard results but they are not trivial to prove. One way to obtain such results is by using the abstract theory of analytic semigroups of operators, see [14, 18, 40].) In ecological applications the function $f(u)$ might be logistic ($f(u) = r(1-[u/K])u$) or might be negative for u small or large but positive for intermediate values, reflecting an Allee effect. Other forms are also possible. The model (3.24) could be generalized by letting the reaction term depend on x or x and t, by adding advection, by replacing the constant diffusion with a variable diffusion rate (leading to a term of the form $\nabla \cdot D(x)\nabla u$), and in various other ways.

Reaction–diffusion systems arise when models for interacting species such as (3.2) or (3.3) are extended by allowing the species to disperse by diffusion. For example, if u and v are densities of competing species, a reaction–diffusion model for them might be

$$\frac{\partial u}{\partial t} = D_u \nabla^2 u + [A - Bu - Cv]u$$

$$\frac{\partial v}{\partial t} = D_v \nabla^2 v + [E - Fu - Gv]v \qquad \text{on } \Omega \times (0,\infty),$$

(3.25)

$$\alpha\frac{\partial u}{\partial \mathbf{n}} + (1 - \alpha)u = \beta\frac{\partial v}{\partial \mathbf{n}} + (1 - \beta)v = 0 \text{ on } \partial\Omega \times (0,\infty),$$

$$u(x,0) = h(x), \ v(x,0) = k(x) \qquad \text{on } \Omega.$$

Again, the model (3.25) will be well posed if $\partial\Omega$ is smooth, as will be similar models with smooth nonlinearities. Adding advection terms or allowing variable coefficients generally does not cause any additional difficulties.

3.1.6 Related Types of Models

Reaction–diffusion models give deterministic descriptions of population densities in continuous space and time. There are a number of related models that give spatially explicit descriptions of populations but which treat space and time in different ways. Those include discrete diffusion models, integrodifference models, and interacting particle systems.

Discrete diffusion models replace continuous space with a collection of discrete locations linked by dispersal. They differ from true metapopulation models because they track population densities rather than patch occupancy, and they are deterministic. A typical discrete diffusion model would describe the population densities $u_i(t)$ on a collection of n patches by a system of ordinary differential equations

$$\frac{du_i}{dt} = \sum_{\substack{j=1 \\ j \neq i}}^{n} D_{ij} u_j - \sum_{\substack{j=1 \\ j \neq i}}^{n} D_{ji} u_i + f_i(u_i),$$

(3.26)

$$i = 1, \ldots, n.$$

The first two terms on the right side of (3.26) describe immigration into and emigration out of the ith patch, while the last term describes population dynamics within the ith patch. An interesting special case of (3.26) occurs when movement occurs only between neighboring patches. In that case the equations in (3.26) for $i = 2, \ldots, n-1$ might take the form

$$\frac{du_i}{dt} = D[u_{i+1} - 2u_i + u_{i-1}] + f_i(u_i).$$

(3.27)

Equation (3.27) is similar to the reaction–diffusion model (3.24), but with a second difference replacing the second derivative terms.

Integrodifference models treat space as continuous but time as discrete. Suppose that a population is at density $u(x, T)$ for $x \in \mathbb{R}$ at time T, reproduces to produce a new density $F(u(x, T))$, then diffuses according to

$$\frac{\partial F}{\partial t} = D \frac{\partial^2 F}{\partial x^2}$$

(3.28)

for t between T and $T + 1$. By using the formula (3.11) we obtain

$$u(x, T + 1) = \frac{1}{\sqrt{4\pi D}} \int_{-\infty}^{\infty} e^{-(x-z)^2/4D} F(u(z, T)) \, dz.$$

(3.29)

Equation (3.29) is a special case of the more general integrodifference model

$$u(x, T + 1) = \int_{-\infty}^{\infty} K(x, z) F(u(z, T)) \, dz.$$

(3.30)

The kernel $K(x, z)$ can be chosen to describe various dispersal processes or can be constructed from data, see [23, 24, 27].

Interacting particle systems typically treat space as a grid or lattice of discrete points, treat time as continuous, and track populations in terms of the number of individuals at each grid point at any given time. They are fully stochastic in that they treat birth, death, and movement to neighboring grid points as random processes that occur at rates that may depend on the number of individuals at a given grid point. If the rules governing the random processes are fairly simple it is possible to use hydrodynamic limits or related methods to derive reaction–diffusion equations as continuum limits of interacting particle systems, see [10].

All of the types of models described here support the same general phenomena as reaction–diffusion models, although they may differ considerably in their detailed behavior. An advantage of using reaction–diffusion models is that they can be analyzed by means of the well-developed theories of ordinary and partial differential equations and dynamical systems. Thus, reaction–diffusion models are well suited to the study of general principles in spatial ecology. They may sometimes represent the best description of a specific situation, but that depends on the details and the scale at which the situation is viewed.

3.2 Solutions of Reaction–Diffusion Equations and Their Basic Properties

3.2.1 Linear Equations, Eigenvalue Problems, and Minimal Patch Size

In general it is not possible to give simple explicit formulas for solutions of reaction–diffusion equations, even in the linear case. However, solutions of linear reaction–diffusion equations can sometimes be expressed in terms of eigenvalues and eigenfunctions of the Laplace operator (or in a single space dimension the second derivative operator) via the method of separation of variables. Consider the one-dimensional model

$$\frac{\partial u}{\partial t} = D\frac{\partial^2 u}{\partial x^2} + ru \quad \text{on } (0, \ell) \times (0, \infty)$$

$$u(0, t) = u(\ell, t) = 0.$$

(3.31)

Equation (3.31) describes the density u of a population inhabiting a one-dimensional patch of length ℓ. The boundary conditions correspond to the assumption that the region outside the patch is lethal to the population. Recall that D is the diffusion rate and r the local intrinsic growth rate of the population. We can solve (3.31) in terms of the eigenvalues and eigenfunctions of the problem

$$\frac{\mathrm{d}^2\phi}{\mathrm{d}x^2} + \lambda\phi = 0 \quad \text{on } 0 < x < \ell$$

$$\phi(0) = \phi(\ell) = 0.$$

(3.32)

It turns out that (3.32) has eigenvalues and eigenfunctions $\lambda_n = -n^2\pi^2/\ell^2$, $\phi_n(x) = \sin(n\pi x/\ell)$, $n = 1, 2, \ldots$. The solution to (3.31) is then given by

$$u(x,t) = \sum_{n=1}^{\infty} c_n\, e^{[r-(Dn^2\pi^2/\ell^2)]t} \sin(n\pi x/\ell), \tag{3.33}$$

where the coefficients c_n depend on the initial density $u(x,0)$.

An important feature of (3.33) is that $u(x,t) \to 0$ as $t \to \infty$ for any initial conditions if $r - (D\pi^2/\ell^2) < 0$. For the model to predict population growth we need $r - D\pi^2/\ell^2 > 0$; in other words

$$\ell > \pi\sqrt{D/r}. \tag{3.34}$$

The condition (3.34) gives the minimal patch size needed to sustain a population. The observation that reaction–diffusion models can predict such a minimal patch size was made by Skellam [44] and Keirstead and Slobodkin [21] in the 1950s. It is worth noting that the quantity $r - D\pi^2/\ell^2$ is the largest eigenvalue, σ_1, of

$$D\frac{d^2\psi}{dx^2} + r\psi = \sigma\psi \quad 0 < x < \ell$$
$$\psi(0) = \psi(\ell) = 0. \tag{3.35}$$

Suppose now that $\Omega \subseteq I\!\!R^n$ is a bounded domain and consider the model

$$\frac{\partial u}{\partial t} = \nabla \cdot D(x)\nabla u + r(x)u \text{ on } \Omega \times (0,\infty)$$
$$\alpha D(x)\frac{\partial u}{\partial \mathbf{n}} + (1-\alpha)u = 0 \text{ on } \partial\Omega \times (0,\infty), \tag{3.36}$$

where $\partial u/\partial \mathbf{n}$ is the outer normal derivative and $\alpha \in [0,1]$ is a parameter specifying the boundary conditions in the model. The associated problem

$$\nabla \cdot D(x)\nabla\psi + r(x)\psi = \sigma\psi \text{ on } \Omega$$
$$\alpha D(x)\frac{\partial u}{\partial \mathbf{n}} + (1-\alpha)u = 0 \text{ on } \partial\Omega \tag{3.37}$$

has a sequence of eigenvalues $\sigma_1 > \sigma_2 \geq \sigma_3 \ldots$ with $\sigma_n \to -\infty$ as $n \to \infty$. Solutions to (3.36) can be written as

$$u(x,t) = \sum_{n=1}^{\infty} c_n\, e^{\sigma_n t}\psi_n(x),$$

where $\psi_n(x)$ is the eigenfunction corresponding to σ_n. In general it is not possible to explicitly calculate the eigenvalues for (3.37), but there is a well-developed theory that can be used to study their dependence on D, r, α, and Ω, see [6]. In the special case where D and r are constants, $\alpha = 0$ (so the exterior of Ω is lethal), and Ω is a dilation of Ω_0 so that $\Omega = \ell\Omega_0 = \{\ell x : x \in \Omega_0\}$ where ℓ is a scale factor, we can separate the biological terms D and r from the geometry of Ω_0 and the scale ℓ. In that case we have $\sigma_n = r - D\lambda_n/\ell^2$, where λ_n is the nth eigenvalue of

$$\nabla^2 \phi + \lambda \phi = 0 \text{ on } \Omega_0$$
$$\phi = 0 \qquad \text{on } \partial\Omega_0.$$

(3.38)

By general theoretical considerations, (3.38) has eigenvalues $0 < \lambda_1 \le \lambda_2 \le \lambda_3 \ldots$, and so in this case we obtain the minimal patch size $\ell > \sqrt{D\lambda_1/r}$. The eigenvalue λ_1 is called the principal eigenvalue for (3.38). Its eigenfunction can be chosen to be positive inside Ω_0. It is interesting to consider how λ_1 depends on the shape of Ω_0. If Ω_0 is a rectangle of width a and length b, then $\lambda_1 = (\pi^2/a^2) + (\pi^2/b^2)$. Hence, for a fixed area ab a long thin rectangle will have a larger principal eigenvalue, and hence a larger minimal size to sustain a population, than would a square. However, λ_1 is relatively insensitive to small scale "wiggles" of the boundary $\partial\Omega_0$, and so λ_1 is not strongly affected by the perimeter to area ratio of Ω_0. The eigenvalue λ_1 is more strongly influenced by the relative amount of "core habitat" in Ω_0 that is far enough from $\partial\Omega_0$ that boundary effects are weak. As in the case of (3.31), the average population growth rate for (3.36) is given by the principal eigenvalue σ_1 of (3.37). For further discussion of eigenvalues and their significance for reaction–diffusion models in ecology see [6]. They will come up again when we consider the stability of equilibria in nonlinear models.

3.2.2 Remark on Existence and Smoothness of Solutions

Although explicit solutions to reaction–diffusion models are usually unavailable, there is a well-developed theory of the existence, uniqueness, and basic properties of solutions. Typical approaches to existence theory are similar to those used for ordinary differential equations such as Picard iteration or the contraction mapping principle (i.e., Banach fixed point theorem.) Recall that to show the existence of solutions to the problem

$$\frac{dy}{dt} = f(t, y), \quad y(0) = y_0$$

(3.39)

we can write (3.39) in integrated form

$$y(t) = y_0 + \int_0^t f(s, y(s)) \, ds$$

(3.40)

and then define a sequence of successive approximations iteratively as

$$y_{n+1}(t) = y_0 + \int_0^t f(s, y_n(s)) \, ds.$$

Under appropriate conditions the sequence $\{y_n(t)\}$ can be shown to converge uniformly to a solution of (3.40) on some interval $[0, T]$. Typical assumptions on $f(t, y)$ are continuity with respect to t and Lipschitz continuity with respect to y. (The function $f(t, y)$ is Lipschitz in y, (uniformly in t) if for every bounded interval $[a, b]$ there is a constant F_0 such that

$$|f(t, y_1) - f(t, y_2)| \leq F_0 |y_1 - y_2| \tag{3.41}$$

whenever $y_1, y_2 \in [a, b]$. Generally the constant depends on a and b. If $\partial f / \partial y$ is continuous then f will be Lipschitz, but there are functions such as $|y|$ that are Lipschitz but not differentiable. Once we have established that (3.40) has a continuous solution $y(t)$ on the interval $[0, T]$ it follows immediately that $y(t)$ is differentiable and hence satisfies (3.39). The estimates giving existence can be adapted to show that the solution is unique. Solutions to (39) may exist only on some finite interval $[0, T_0)$ and then become infinite. An example where that occurs is $dy/dt = y^2$.

The general considerations for reaction–diffusion models are similar. The main difference is that it is not so easy to see that solutions to integrated problems such as (3.40) are actually smooth. The solutions of linear reaction–diffusion equations can be expressed as integrals involving Green's functions. For the problem

$$\frac{\partial u}{\partial t} = D \frac{\partial^2 u}{\partial x^2} \quad \text{on } \mathbb{R} \times (0, \infty)$$

the Green's function is

$$K(x, t) = \frac{1}{\sqrt{4\pi D t}} e^{-x^2 / 4Dt}. \tag{3.42}$$

The solution of

$$\frac{\partial u}{\partial t} = D \frac{\partial^2 u}{\partial x^2} + f(t, x) \text{ on } \mathbb{R} \times (0, \infty)$$

$$u(x, 0) = h(x) \qquad \text{on } \mathbb{R}$$

is given by

$$u(x, t) = \int_{-\infty}^{\infty} K(x - y, t) h(y) \, dy + \int_0^t \int_{-\infty}^{\infty} K(x - y, t - s) f(s, y) \, dy ds \tag{3.43}$$

where K is as in (3.42). It is only sometimes possible to give explicit formulas for Green's functions, but their existence can be shown by either classical methods [14] or the theory of analytic semigroups of operators [15].

Suppose $\Omega \subseteq \mathbb{R}^n$ is a domain and that $G(x, y, t)$ is the Green's function for

$$\frac{\partial u}{\partial t} = \nabla \cdot D(x) \nabla u \qquad \text{on } \Omega \times (0, \infty)$$

$$\alpha D(x) \frac{\partial u}{\partial \mathbf{n}} + (1 - \alpha) u = 0 \text{ on } \partial \Omega \times (0, \infty)$$

$$u(x, 0) = h(x) \qquad \qquad \text{on } \Omega.$$

For the reaction–diffusion model

$$\frac{\partial u}{\partial t} = \nabla \cdot D(x)\nabla u + f(x,t,u) \quad \text{on } \Omega \times (0,\infty)$$

$$\alpha D(x)\frac{\partial u}{\partial \mathbf{n}} + (1-\alpha)u = 0 \quad \text{on } \partial\Omega \times (0,\infty) \qquad (3.44)$$

$$u(x,0) = h(x) \quad \text{on } \Omega$$

the formula analogous to (3.40) is

$$u(x,t) = \int_\Omega G(x,y,t)h(y)\,\mathrm{d}y + \int_0^t \int_\Omega G(x,y,t-s)f(y,s,u(y,s))\,\mathrm{d}y\,\mathrm{d}s. \quad (3.45)$$

As in (3.40) we would normally want to assume that f is Lipschitz in u to obtain existence of solutions to (45). In that case we could proceed by using Picard iteration or the contraction mapping principle. In the case where Ω is unbounded we would normally require u and the coefficients and data for (3.44) to be bounded.

A technical problem with (3.45) relative to (3.44) is that even if f depends only on x and t (so the equation (3.44) is linear) and f is continuous in x and t, the expression on the right in (3.45) is not necessarily continuously differentiable in t or twice continuously differentiable in the components of x, so that u may not represent a classical solution of (3.44). On the other hand, the Green's function $K(x,t)$ in (3.43) is very smooth and decays rapidly as $|x| \to \infty$ for $t > 0$, and so in the case $f(x,t) = 0$ the solution given by (3.43) is smooth for $t > 0$. These observations are the starting point for regularity theory for reaction–diffusion equations and parabolic equations in general.

To describe the regularity properties of reaction–diffusion equations we need to introduce the idea of Hölder continuity. A function $g(x)$ on a closed domain $\bar\Omega \subseteq I\!\!R^n$ is uniformly Hölder continuous with exponent $\gamma \in (0,1)$ if there is constant G_0 such that

$$|g(x_1) - g(x_2)| < G_0|x_1 - x_2|^\gamma$$

for $x_1, x_2 \in \bar\Omega$. (Compare with the definition of Lipschitz continuity in (3.41), where $\gamma = 1$.) A function on $\bar\Omega \times [0,T]$ whose derivatives of order k or less in x are bounded and Hölder continuous with exponent γ and whose derivatives of order ℓ or less in t are bounded and Hölder continuous with exponent δ is said to belong to the space $C^{k+\gamma,\ell+\delta}(\bar\Omega \times [0,T])$. Such functions that depend only on x are said to belong to $C^{k+\gamma}(\bar\Omega)$. The space $C^{k+\gamma}(\bar\Omega)$ has the norm

$$\|u\|_{k+\gamma} = \sup_{x\in\bar\Omega}|u(x)| + \sum_{|\beta|\le k} \sup_{x\in\bar\Omega}|\partial^\beta u| + \sum_{|\beta|=k} \sup_{x_1,x_2\in\bar\Omega} \frac{|\partial^\beta u(x_1) - \partial^\beta u(x_2)|}{|x_1 - x_2|^\gamma},$$

where $\{\partial^\beta u : |\beta| = m\}$ refers to the set of all derivatives of u of order m. The norm for $C^{k+\gamma,\ell+\delta}(\bar\Omega \times [0,T])$ includes the analogous terms involving

t derivatives of order ℓ or less. Under those norms $C^{k+\gamma}(\bar{\Omega})$ and $C^{k+\gamma,\ell+\delta}$ ($\bar{\Omega} \times [0,T]$) are Banach spaces. Suppose that in (3.44) the function f depends only on x and t, Ω is bounded with $\partial\Omega$ smooth, and $D(x) \geq D_0 > 0$ with $D(x) \in C^{1+\gamma}(\bar{\Omega})$. If $f(x,t) \in C^{\gamma,\gamma/2}(\bar{\Omega} \times [0,T])$ then for any $\epsilon > 0$ the "solution" to (3.44) defined by (3.45) will belong to $C^{2+\gamma,1+\gamma/2}(\bar{\Omega} \times [\epsilon,T])$, so it will indeed be a solution to (3.44). (Under suitable conditions on $h(x)$ we can take $\epsilon = 0$.) Similar results are available in considerably more generality, in particular for models with advection and/or time dependent coefficients. They are important in two ways. First, they are needed to verify that solutions of (45) or other generalized solutions of (3.44) are actually classical solutions. Second, regularity theory can be used together with ideas such as the Arzela–Ascoli theorem to show that bounded sequences of solutions have convergent subsequences, or that solution trajectories which are bounded for all time have compact closures in the appropriate function spaces, or various other results related to compactnesss. Compactness results turn out to be essential for the application of dynamical systems theory and some forms of bifurcation theory to reaction–diffusion equations and systems. (The theory of reaction–diffusion equations as dynamical systems is treated in [18].)

The important ideas in this section are that under reasonable smooth-ness assumptions on the coefficients and underlying spatial domain, reaction–diffusion equations have well defined unique solutions for at least some finite time interval. The constructions used to show existence can be applied as long as solutions remain finite, but in some cases solutions may grow to infinity in finite time. To obtain the existence of classical solutions requires informa-tion about the smoothness of genralized solutions, that is, regularity theory. Regularity theory describes how reaction–diffusion equations smooth out ini-tial data, among other things. It can be used to establish compactness results needed for various sorts of analysis of reaction–diffusion models.

A secondary idea introduced in this section is the notion that the regularity of solutions can be described in terms of function spaces, specifically spaces of Hölder continuous functions. Also, differential operators and their inverses can be viewed as maps between functions spaces, which allows the application of deep results from functional analysis. Hölder spaces are only some of the relevant function spaces for the study of partial differential equations.

Various aspects of existence and regularity theory for partial differential equations including reaction–diffusion equations are discussed in [14, 15, 18, 25, 38, 40, 46]. The theory extends to systems of the form

$$\frac{\partial u_i}{\partial t} = \nabla \cdot [D_i(x,t)\nabla u_i + \mathbf{b}_i(x,t)u_i] + f_i(x,t,u_1,\ldots,u_n)$$

in $\Omega \times (0,T]$,

$$\alpha_i(x)\frac{\partial u_i}{\partial \mathbf{n}} + \beta_i(x)u_i = 0 \quad \text{on } \partial\Omega \times (0,T], \quad i = 1,\ldots,n,$$

if $D_i(x,t) \geq D_0 > 0$, $\alpha_i \geq 0$, $\beta_i \geq 0$, and $\alpha_i + \beta_i > 0$, $i = 1, \ldots, n$, and the coefficients and $\partial \Omega$ are sufficiently smooth. The main limitation of existence theory per se is that it is local in time. As long as $|u(x,t)|$ or $|\mathbf{u}(x,t)|$ remains bounded the solution can be continued forward in time, but it is possible that $|u(x,t)| \to \infty$ as $t \uparrow T$ for some T. In the case of scalar equations this problem can often be addressed by using results from the next section; for systems it remains somewhat more delicate.

3.2.3 Maximum Principles and Comparison Theorems

For the ordinary differential equation $y' = f(y)$ with f smooth, the initial data $y(0) = y_0$ uniquely determines a solution, and so if $y_1(t)$ and $y_2(t)$ are solutions with $y_1(0) \geq y_2(0)$ then either $y_1(0) = y_2(0)$ and hence $y_1(t) = y_2(t)$ or $y_1(0) > y_2(0)$ and we must have $y_1(t) > y_2(t)$. For the logistic equation $y' = r[1 - (y/K)]y$, it follows that if $0 < y(0) < 1$ then $0 < y(t) < 1$ for all $t > 0$ since $y \equiv 0$ and $y \equiv 1$ are solutions. The situation for reaction–diffusion equations seems more complicated since solutions depend on both time and space, but it turns out that something similar is often true of their solutions. The basis for such results for reaction–diffusion equations is called the maximum principle. The maximum principle can be stated as a theorem in various ways. (Sometimes it is easier to adapt the proof to answer a specific question than it is to use some particular version of the theorem.) One version is as follows:

Theorem 1. *Assume that $u(x,t)$ is a classical solution to the differential inequality*

$$\frac{\partial u}{\partial t} \geq D(x,t)\nabla^2 u + \mathbf{b}(x,t) \cdot \nabla u + c(x,t)u \ \text{ on } \ \Omega \times (0,T), \qquad (3.46)$$

that $u(x,t)$ is continuous on $\bar{\Omega} \times [0,T]$, and if Ω is unbounded assume also that u is bounded on $\bar{\Omega} \times [0,T]$. Assume that the coefficients of (3.46) are continuous, and if Ω is unbounded assume that they are also bounded. Assume $D \geq D_0 > 0$. Then

(i) *If $u \geq 0$ on $(\Omega \times \{0\}) \cup (\partial\Omega \times [0,T])$ then $u \geq 0$ on $\bar{\Omega} \times [0,T]$.*
    Assume further that Ω is connected. Then
(ii) *If $u \geq 0$ on $\bar{\Omega} \times [0,T]$ and $u(x_0,t_0) = 0$ for some $x_0 \in \Omega$ and $t_0 > 0$, then $u(x,t) \equiv 0$ for $x \in \bar{\Omega}$, $t \in [0,t_0]$.*
(iii) *If $u \geq 0$ on $\bar{\Omega} \times [0,T]$ and $u(x_0,t_0) = 0$ for some $x_0 \in \partial\Omega$ and $t_0 > 0$ then either $\dfrac{\partial u}{\partial \mathbf{n}} < 0$ at (x_0,t_0) or $u \equiv 0$ for $x \in \bar{\Omega}$, $t \in [0,t_0]$.*

The proof of (i) is fairly direct; the proofs of (ii) and (iii) are more complicated. We will sketch the proof of (i). For (ii) and (iii) see [14, 41].

First an observation: We can assume that $c \geq 0$ or $c \leq 0$ without loss of generality. Here is why: if u satisfies (3.2) then $w = e^{Kt}u$ satisfies

$$\frac{\partial w}{\partial t} \geq D(x,t)\nabla w^2 + \mathbf{b}(x,t) \cdot \nabla w + [c(x,t) + K]w.$$

Since c is bounded we can make $c + K$ positive or negative by choosing K correctly, and $w \geq 0$ if and only if $u \geq 0$, $w = 0$ if and only if $u = 0$.

Proof of (i) for Ω bounded:

First, suppose that v is a function satisfying

$$\frac{\partial v}{\partial t} > D(x,t)\nabla^2 v + \mathbf{b}(x,t) \cdot \nabla v + c(x,t)v \quad \text{in } \Omega \times (0,T],$$

$$v > 0 \quad \text{on } (\Omega \times \{0\}) \cup (\partial\Omega \times [0,T]). \tag{3.47}$$

If $v(x,t) \leq 0$ somewhere in $\bar{\Omega} \times [0,T]$ let $t_0 = \min\{t : v(x,t) \leq 0 \text{ for some } x \in \bar{\Omega}\}$. (Since $v(x,t)$ is differentiable and hence continuous the set $\{(x,t) : v(x,t) \leq 0\}$ is closed so the function $g(t) = t$ attains a minimum on it if it is nonempty.) If $v(x_1,t_0) < 0$ for some $x_1 \in \bar{\Omega}$ then $v(x_1,t) < 0$ for $t < t_0$, $t \approx t_0$ so we must have $v(x,t_0) \geq 0$, $v(x_0,t_0) = 0$ for some $x_0 \in \bar{\Omega}$. By the boundary condition x_0 cannot belong to $\partial\Omega$, so (3.47) must hold at (x_0,t_0). Also, $v(x,t) > 0$ for $t < t_0$, so $v(x_0,t) > 0 = v(x_0,t_0)$ for $t < t_0$, so $\partial v/\partial t \leq 0$ at (x_0,t_0). Since $x_0 \in \Omega$ and $v(x,t_0) \geq 0$, $v(x_0,t_0) = 0$, the function $v(x,t_0)$ has a local minimum at $x = x_0$, so $\partial v/\partial x_i = 0$ and $\partial^2 v/\partial x_i^2 \geq 0$ at (x_0,t_0). Using these observations in (3.47) implies that at (x_0,t_0)

$$0 \geq \frac{\partial v}{\partial t} > D\nabla^2 v + \mathbf{b} \cdot \nabla v + cv \geq 0. \tag{3.48}$$

Since (3.48) implies $0 > 0$, which is a contradiction, we cannot have $v \leq 0$ anywhere in $\bar{\Omega} \times [0,T]$, so $v > 0$.

Now consider the case where u satisfies (3.46). (We can assume $c \leq 0$ in (3.46) without loss of generality.) For any $\epsilon > 0$ we can let $v = u + \epsilon e^t$. We have $v > 0$ on $(\Omega \times \{0\}) \cup (\partial\Omega \times [0,T])$ because $u \geq 0$ on that set. Also, $\nabla v = \nabla u$, so we have

$$\frac{\partial v}{\partial t} = \epsilon e^t + \frac{\partial u}{\partial t}$$

$$\geq \epsilon e^t + D\nabla^2 v + \mathbf{b} \cdot \nabla v + c(v - \epsilon e^t)$$

$$= D\nabla^2 v + \mathbf{b} \cdot \nabla v + cv + \epsilon(1 - c)e^t$$

so $\dfrac{\partial v}{\partial t} > D\nabla^2 v + \mathbf{b} \cdot \nabla v + cv$ since $c \leq 0$. It follows that $v > 0$, so $u > -\epsilon e^t$. Since $\epsilon > 0$ was arbitrary, we have $u \geq 0$, which completes the proof.

Remark. Suppose that instead of the boundary condition $u \geq 0$ in (i) we have

$$\frac{\partial u}{\partial n} + \beta(x)u \geq 0 \quad \text{on } \partial\Omega \times (0,T]. \tag{3.49}$$

Assume that $\beta(x) \geq 0$ and that Ω and the coefficients of (3.46) are as in Theorem 1. (If Ω is unbounded assume that u, the coefficients of (3.46), and $\beta(x)$ are bounded.) Then $u \geq 0$ in $\bar{\Omega} \times [0,T]$. This can be seen by using part (iii) of Theorem 1; we will sketch the argument here.

If $u < 0$ somewhere on $\bar{\Omega} \times (0,T]$ then u has a negative minimum $-M$. Let $w = e^{Kt}u$ with K chosen so that $c + K < 0$; then $w < 0$ somewhere on $\bar{\Omega} \times (0,T]$ and w satisfies

$$\frac{\partial w}{\partial t} \geq \nabla \cdot D\nabla w + \mathbf{b} \cdot \nabla w + (c+K)w \quad \text{in } \Omega \times (0,T]$$

with $w(x,0) \geq 0$. Since $w < 0$ somewhere, w must have a negative minimum $-\tilde{M}$. Let $v = w + \tilde{M}$. Then the minimum of v on $\bar{\Omega} \times [0,T]$ is 0. Also, we have $v \geq 0$ on $\bar{\Omega} \times [0,T]$ and $v_t = w_t \geq \nabla \cdot D\nabla w + \mathbf{b} \cdot \nabla w + (c+K)w = \nabla \cdot D\nabla v + \mathbf{b} \cdot \nabla v + (c+K)v - (c+K)\tilde{M}$ since $\nabla v = \nabla w$. Since $-(c+K)\tilde{M} > 0$, v satisfies (3.47). Thus, as in the proof of part (i), Theorem 1, v cannot attain its minimum in $\Omega \times (0,T]$. It follows that v must have its minimum, $v = 0$, at some point of $\partial\Omega \times (0,T]$. At such a point we have by (3.50)

$$\frac{\partial v}{\partial \mathbf{n}} = \frac{\partial w}{\partial \mathbf{n}} \geq -\beta w = -\beta v + \beta\tilde{M} = \beta\tilde{M} \geq 0. \tag{3.50}$$

On the other hand, by part (iii) of Theorem 1, either $v \equiv 0$ in $\bar{\Omega} \times [0,T]$ or $\dfrac{\partial v}{\partial \mathbf{n}} < 0$ at the point on $\partial\Omega \times (0,T]$ where v has its minimum. By (3.50), $\dfrac{\partial v}{\partial \mathbf{n}} < 0$ is impossible, but because v cannot attain its minimum in $\Omega \times (0,T]$, $v \equiv 0$ is also impossible. We have arrived at a contradiction, so our assumption that u was negative somewhere in $\bar{\Omega} \times [0,T]$ cannot be valid. Hence we must have $u \geq 0$ in $\bar{\Omega} \times [0,T]$.

If the coefficients of (3.46) do not depend on t then the equilibrium model for (3.46) will have a maximum principle provided $c(x) \leq 0$. In the equilibrium case that condition is a genuine restriction. (If $c(x) \equiv 0$ we may also need $\beta(x) > 0$ for some $x \in \partial\Omega$.). For example, if $u(x)$ satisfies

$$0 \geq D(x)\nabla^2 u + \mathbf{b}(x) \cdot \nabla u + c(x)u \quad \text{on } \Omega$$

with $c(x) \leq 0$ and $u(x) \geq 0$ on $\partial\Omega$, then $u \geq 0$ on $\bar{\Omega}$ as in (i) of Theorem 1.

One of the main applications of the maximum principle is for comparisons between solutions of differential equations and inequalities. We will turn to that topic next. We have

Theorem 2. *Suppose that Ω and the coefficients $D, \mathbf{b}$, and (if relevant) β are as in (3.46) and (3.49), respectively, that $f(x,t,u)$ is Lipschitz in u, and Ω is connected. Suppose further that*

$$\frac{\partial \bar{u}}{\partial t} - D(x,t)\nabla^2 \bar{u} - \mathbf{b}(x,t) \cdot \nabla \bar{u} - f(x,t,\bar{u})$$

$$\geq \frac{\partial \underline{u}}{\partial t} - D(x,t)\nabla^2 \underline{u} - \mathbf{b}(x,t) \cdot \nabla \underline{u} - f(x,t,\underline{u}) \quad on \ \Omega \times (0,T],$$

$\bar{u}(x,0) \geq \underline{u}(x,0) \ on \ \Omega,$
and either   (3.51)
$\bar{u} \geq \underline{u} \ on \ \partial\Omega \times (0,T]$
or

$$\frac{\partial \bar{u}}{\partial \mathbf{n}} + \beta(x)\bar{u} \geq \frac{\partial \underline{u}}{\partial \mathbf{n}} + \beta(x)\underline{u} \ on \ \partial\Omega \times (0,T].$$

(If Ω is unbounded assume that $\bar{u}, \underline{u}, d, \mathbf{b},$ and β are bounded.)
Then

$$\bar{u} \geq \underline{u} \ in \ \bar{\Omega} \times [0,T],$$

and for any $t_0 \in (0,T]$ we have $\bar{u}(x,t_0) > \underline{u}(x,t_0)$ on Ω or $\bar{u} \equiv \underline{u}$ on $\bar{\Omega} \times [0,t_0]$.

Proof. Let $w = \bar{u} - \underline{u}$. Since f is Lipschitz in u the expression $\dfrac{f(x,t,\bar{u}) - f(x,t,\underline{u})}{\bar{u} - \underline{u}}$
is bounded. We have from (3.51) that

$$\frac{\partial w}{\partial t} = D(x,t)\nabla^2 w - \mathbf{b}(x,t) \cdot \nabla w - \left[\frac{f(x,t,\bar{u}) - f(x,t,\underline{u})}{\bar{u} - \underline{u}} \right] w \geq 0$$

$$in \ \Omega \times (0,T].$$

Since f is Lipschitz in u and w satisfies boundary conditions arising from those in (3.51), we may apply Theorem 1 to w to obtain the conclusions of Theorem 2. □

Theorem 2 implies that the model

$$\frac{\partial u}{\partial t} = D(x)\nabla^2 u + \mathbf{b}(x) \cdot \nabla u + f(x,u) \ in \ \Omega \times (0,T],$$

$$\alpha \frac{\partial u}{\partial \mathbf{n}} + \beta u = 0 \ on \ \partial\Omega \times (0,T] \tag{3.52}$$

is order preserving in the sense that if $\bar{u}(x,t)$ and $\underline{u}(x,t)$ are solutions with $\bar{u}(x,0) \geq \underline{u}(x,0)$, then $\bar{u}(x,t) \geq \underline{u}(x,t)$ for $t \in (0,T]$. There is a formal theory of order preserving dynamical systems, see [19, 45]. A key feature of the theory is that roughly speaking the long-term dynamics of such systems are mostly determined by their equilibria in the sense that bounded global solutions generically approach the set of equilibria as $t \to \infty$. In particular, periodic orbits may exist but if so they are unstable. Some but not all reaction–diffusion systems admit maximum principles and comparison theorems. For the system

$$\frac{\partial u_i}{\partial t} = D_i \nabla^2 u_i + f_i(x, t, \mathbf{u}) \text{ on } \Omega \times (0, T], \quad i = 1, \ldots, n,$$

$$\alpha_i \frac{\partial u_i}{\partial \mathbf{n}} + \beta_i u_i = 0 \qquad \text{ on } \partial\Omega \times (0, T], \quad i = 1, \ldots, n \tag{3.53}$$

with our usual assumptions on the coefficients, the componentwise analogue of Theorem 2 holds provided the coupling terms f_i satisfy

$$\frac{\partial f_i}{\partial u_j} \geq 0 \text{ for } i \neq j. \tag{3.54}$$

Systems satisfying (3.54) are known as cooperative or quasi-monotone. Unless the system (3.53) satisfies (3.54) it will generally not admit a comparison principle; however, there are some cases where a model that does not satisfy (3.54) can be made to do so by a change of variables. The most important example in theoretical ecology is the case of models for two competing species. Those typically have with $\partial f_i / \partial u_j \leq 0$ for $i \neq j$, at least for $u_1, u_2 \geq 0$. For such systems the ordering $\mathbf{u} \geq \mathbf{v} \Leftrightarrow u_1 \geq v_1, \; u_2 \leq v_2$ is typically preserved. Models for three or more competitors do not usually admit comparison principles; neither do most predator–prey models. However, it is still sometimes possible to use related ideas to study them. One approach is to apply maximum principles to one equation at a time, see [5, 7]. Another is to embed the system into a larger system that is cooperative, see [8].

A different sort of application of maximum principles to systems of equations is to use them to extend the notion of invariant sets from the theory of ordinary differential equations to reaction–diffusion systems. For the system

$$\frac{du_i}{dt} = f_i(\mathbf{u}),$$

a region $S \subseteq \mathbb{R}^n$ is invariant forward in time if the vector field $\mathbf{f} = (f_1, \ldots, f_n)$ never points out of S on ∂S. The same is true for the system (3.53) if $D_i = D_j$ for all i, j and S is convex, or for $D_i > 0$ if S is a rectangle. More general formulations are possible, but some restrictions on S and the differential operators are needed. Results on invariance are derived and applied to systems, including predator–prey models in [46].

Maximum principles for various types of equations and systems are discussed in [41]. Topics related to maximum principles and comparison theorems are discussed in [6, 7, 11, 19, 25, 38, 45].

3.2.4 Sub- and Supersolutions and Stability

A common approach to the application of comparison principles is the use of auxilliary functions that satisfy differential inequalities related to the model. If $\underline{u}(x, t)$ satisfies the inequalities

$$\frac{\partial \underline{u}}{\partial t} \leq D(x,t)\nabla^2 \underline{u} + \mathbf{b}(x) \cdot \nabla \underline{u} + f(x,t,\underline{u}) \quad \text{in } \Omega \times (0,T]$$

(3.55)

$$\alpha \frac{\partial \underline{u}}{\partial \mathbf{n}} + \beta \underline{u} \leq 0 \quad \text{on } \partial\Omega \times (0,T],$$

then $\underline{u}$ is said to be a subsolution or lower solution to the equation

$$\frac{\partial u}{\partial t} = D(x,t)\nabla^2 u + \mathbf{b}(x,t) \cdot \nabla u + f(x,t,u) \quad \text{in } \Omega \times (0,T]$$

(3.56)

$$\alpha \frac{\partial u}{\partial \mathbf{n}} + \beta u = 0 \quad \text{on } \partial\Omega \times (0,T].$$

It follows immediately from Theorem 2 that if u is a solution to (56) and $\underline{u}$ is a subsolution with the same initial data u then $\underline{u} \leq u$ for $t \in (0,T]$, and either $\underline{u} < u$ or $\underline{u} \equiv u$. A supersolution or upper solution of (3.56) would satisfy inequalities obtained by reversing those in (3.55). If $D, \mathbf{b}$, and f are independent of t, we can also define sub- and supersolutions to the equilibrium problem for (3.56). A subsolution $\underline{u}$ would satisfy

$$0 \leq D(x)\nabla^2 \underline{u} + \mathbf{b}(x) \cdot \nabla \underline{u} + f(x,\underline{u}) \quad \text{on } \Omega$$

(3.57)

$$\alpha \frac{\partial \underline{u}}{\partial \mathbf{n}} + \beta \underline{u} \leq 0 \qquad\qquad \text{on } \partial\Omega,$$

while a supersolution would satisfy the reverse inequalities.

Sub- and supersolutions can be used to show the existence and stability or instability of equilibria and to give bounds on solutions in reaction–diffusion models. The following result illustrates some of those themes.

Theorem 3. *Suppose that the coefficients of an autonomous reaction–diffusion equation of the form (3.53) are smooth. Suppose also that u is a solution of equation (3.53) with $u(x,0) = \underline{u}(x)$, where $\underline{u}$ is a subsolution to the corresponding equilibrium problem, i.e., $\underline{u}$ satisfies (3.57). If $\underline{u}$ is not an equilibrium of (3.53) then $u(x,z)$ is increasing in t, and if $u(x,t)$ is bounded from above then $u(x,t) \to u^*(x)$ as $t \to \infty$, where u^* is an equilibrium of (3.53). Similarly if $u(x,0) = \bar{u}(x)$ where $\bar{u}$ is a supersolution to the equilibrium problem then unless $\bar{u}$ is an equilibrium of (53), $u(x,t)$ is decreasing, and $u(x,t)$ converges to an equilibrium if u is bounded from below. (If $\underline{u} \leq \bar{u}$ it follows that there exists an equilibrium u^* with $\underline{u} \leq u^* \leq \bar{u}$.) (As in Theorem 2, if Ω is unbounded, we require the coefficients and sub- or supersolutions to be bounded.)*

Proof (Sketch). Since $\underline{u}(x)$ satisfies (3.57) and does not depend on t, $\underline{u}$ is a subsolution of (3.53). Since u and $\underline{u}$ have the same initial data, $u(x,t) \geq \underline{u}(x)$ for $t \geq 0$ as long as $u(x,t)$ exists. For any $\delta > 0$ we have $u(x,\delta) \geq u(x,0) = \underline{u}(x)$, with strict inequality unless $\underline{u}$ is an equilibrium of (3.53). Let $v = u(x,t+\delta)$. We have $v(x,0) = u(x,\delta) \geq u(x,0)$ and v satisfies (3.53), so by Theorem 2 we have $v(x,t) \geq u(x,t)$ so that $u(x,t+\delta) \geq u(x,t)$. Thus, u is increasing in t. If $u(x,t)$ is bounded from above then for each $x \in \bar{\Omega}$ we have $u(x,t) \to u^*(x)$ for some function $u^*(x)$. To see that u^* is an equilibrium of (3.53), note that the regularity theory for parabolic equations implies that the derivatives of u of order up to two in x and order one in t are uniformly bounded and equicontinuous, and so by the Arzela–Ascoli theorem we may choose a sequence $u(x,t_n)$ with $t_n \to \infty$ such that those derivatives converge uniformly. Since $u(x,t) \to u^*(x)$, u^* must be a solution of (3.53); but u^* does not depend on t so u^* is an equilibrium. $\square$

Remark. A result of this type was obtained in [3]. Many results about sub- and supersolutions for reaction–diffusion equations and systems are discussed in [19, 25, 38]. Related results can be formulated for order preserving dynamical systems in general; see [45].

To apply Theorem 3, or more generally to obtain information from Theorem 2, we need to find suitable sub- and supersolutions or other expressions that can be used for purposes of comparison. Often those can be constructed from solutions to related equations. Consider the diffusive logistic equation on a bounded domain Ω:

$$\frac{\partial u}{\partial t} = D\nabla^2 u + ru(1 - [u/K]) \text{ on } \Omega \times (0, \infty)$$

$$\alpha \frac{\partial u}{\partial \mathbf{n}} + \beta u = 0 \text{ on } \partial\Omega \times (0, \infty).$$

$$(3.58)$$

Note that $u \equiv 0$ is a solution, so if $u(x,t)$ is a solution with $u(x,0) \geq 0$ then $u(x,t) \geq 0$ as long as $u(x,t)$ exists. Also, any constant $U > K$ satisfies

$$0 \geq D\nabla^2 U + rU(1 - [U/K])$$

$$\alpha \frac{\partial U}{\partial \mathbf{n}} + \beta U \geq 0$$

so U is a supersolution to the equilibrium problem for (3.58). If $u(x,t)$ is a solution of (3.58) with $u(x,0) = U$ then by Theorem 3, $u(x,t)$ is decreasing. Since we can choose U to be as large as we like, it follows that all solutions of (3.58) with bounded nonnegative initial data remain bounded and nonnegative, and thus exist for all $t > 0$. Furthermore, since solutions with $u(x,0) = U > K$ decrease toward an equilibrium, they must either approach zero or converge to some other positive equilibrium as $t \to \infty$. To obtain further information we need to consider the stability of the zero solution. The linearized problem for (3.58) around $u \equiv 0$ is

$$\frac{\partial u}{\partial t} = D\nabla^2 u + ru \text{ in } \Omega \times (0, \infty)$$

$$\alpha \frac{\partial u}{\partial \mathbf{n}} + \beta u = 0 \quad \text{on } \partial\Omega \times (0, \infty).$$

$$(3.59)$$

The equilibrium $u \equiv 0$ is linearly stable if the largest eigenvalue, σ_1, is negative in the eigenvalue problem

$$D\nabla^2\psi + r\psi = \sigma\psi \text{ in } \Omega$$

$$\alpha\frac{\partial\psi}{\partial\mathbf{n}} + \beta\psi = 0 \quad \text{on } \partial\Omega.$$

(3.60)

If $\sigma_1 > 0$ then $\bar{u} \equiv 0$ is unstable.

Let λ_1 be the smallest eigenvalue of the problem

$$\nabla^2\phi + \lambda\phi = 0 \text{ in } \Omega$$

$$\alpha\frac{\partial\phi}{\partial\mathbf{n}} + \beta\phi = 0 \text{ on } \partial\Omega.$$

(3.61)

Then $\lambda_1 \geq 0$, $\lambda_1 > 0$ if $\beta \neq 0$, $\sigma_1 = r - D\lambda_1$, and we may choose $\psi = \phi =$ the eigenfunction for λ_1 to be positive in Ω. (These are standard results about eigenvalues and eigenfunctions of the Laplace operator and similar elliptic operators.) Suppose that $\sigma_1 > 0$ in (60). Let $\underline{u} = \epsilon\psi_1$ for some fixed eigenfunction $\psi_1 > 0$ for (3.60) and some $\epsilon > 0$. We have

$$D\nabla^2\underline{u} + r\underline{u}(1 - [\underline{u}/K]) = \epsilon(D\nabla^2\psi_1 + r\psi_1 - [r\epsilon/K]\psi_1^2)$$

$$= \epsilon(\sigma_1\psi_1 - [r\epsilon/K]\psi_1^2)$$

$$= \epsilon\psi_1(\sigma_1 - [r\epsilon/K]\psi_1) > 0$$

for $\epsilon > 0$ sufficiently small. Thus, if $\sigma_1 = r - D\lambda_1 > 0$, i.e., $r/D > \lambda_1$, then $\underline{u} = \epsilon\psi_1 > 0$ is a subsolution to the equilibrium problem for (3.58). Hence, if $\sigma_1 > 0$, we can find sub- and supersolutions $0 < \underline{u} = \epsilon\psi_1 < U = \bar{u}$ for the equilibrium problem for (3.58), and so in that case (58) has a positive equilibrium u^*. Also, $\epsilon > 0$ can be chosen to be arbitrarily small, so any positive solution can be bounded below by a solution with $u(x,0) = \epsilon\psi_1$, which increases toward a positive equilibrium. Thus, the model predicts uniform persistence of the population in this case. (It turns out that the positive equilibrium of (3.58) is unique when it exists, but that requires a separate argument. See [6], Ch. 3 for further discussion of this and related points). What if $\sigma_1 < 0$ so that $u \equiv 0$ is stable? It turns out that there is no positive equilibrium u^* in that case. One of the consequences of the divergence theorem is Green's formula (see [31, 47] for example)

$$\int_\Omega (v\nabla^2 w - w\nabla^2 v) \, dx = \int_{\partial\Omega} \left(v\frac{\partial w}{\partial\mathbf{n}} - w\frac{\partial v}{\partial\mathbf{n}}\right) dS.$$

(3.62)

If u^* is an equilibrium of (3.58) then u^* and ψ_1 satisfy the same boundary conditions, so

$$0 = D \int_{\partial\Omega} \left(u^* \frac{\partial \psi_1}{\partial \mathbf{n}} - \psi_1 \frac{\partial u^*}{\partial \mathbf{n}} \right) dS$$

$$= \int_\Omega [u^*(D\nabla^2\psi_1) - \psi_1(D\nabla^2 u^*)] \, dx$$

$$= \int_\Omega (u^*[-r\psi_1 + \sigma_1\psi_1] - \psi_1[-ru^* + (r/K)u^{*2}]) \, dx \qquad (3.63)$$

$$= \int_\Omega [\sigma_1 - (r/K)u^*]u^*\psi_1 \, dx.$$

If $u^* > 0$ then the last integral is negative for $\sigma_1 \leq 0$, which yields a contradiction. Hence, (3.58) cannot have a positive equilibrium in that case. However, solutions with $u(x,0) = U > K$ must decrease toward some equilibrium and must remain nonnegative. It follows that $u \equiv 0$ is globally stable among nonnegative solutions of (3.58) when $\sigma_1 \leq 0$ in (3.60). We have shown that positive solutions of (3.58) exist and remain bounded for all $t > 0$. If $r/D > \lambda_1$, where λ_1 is defined in (3.61), then $\sigma_1 > 0$ in (3.60), (3.58) has at least one positive equilibrium u^*, and $u \equiv 0$ is unstable in the sense that solutions with positive initial data are bounded below by solutions that increase toward a positive equilibrium. If $r/D \leq \lambda_1$ then $\sigma_1 \leq 0$ and all positive solutions approach zero as $t \to \infty$. The condition $r/D > \lambda_1$ is in effect a requirement that Ω be sufficiently large, since λ_1 generally decreases if Ω is enlarged. (Recall the discussion at the end of Sect. 2.1. Under Dirichlet boundary conditions, λ_1 scales like $1/\ell^2$ if Ω is rescaled to $\ell\Omega = \{\ell x : x \in \Omega\}$.)

The idea of sub- and supersolutions can be used to obtain results along the lines of Theorem 3 for many reaction–diffusion systems. Specifically, models for two competing species such as (3.25) can be treated in a similar way, see [6, 19, 25, 38]. A key element in the analysis is the order-preserving property arising from the comparison principle, but it is sometimes possible to use related ideas to treat systems without that property, see [7, 25, 38]. In a different direction, Theorem 3 can be used to study reaction–diffusion models on unbounded domains. Such models can sometimes support solutions such as traveling wavefronts that propagate through space, and hence they can be used to describe biological invasions. We will examine these ideas in some depth in the next section.

3.3 Traveling Waves

3.3.1 Wavefronts

Reaction–diffusion equations have been used extensively as models for biological invasions. In that setting the spatial domain is usually taken to be $\mathbb{R}$. It turns out that reaction–diffusion equations on $\mathbb{R}$ can support traveling wavefronts, that is, positive solutions of the form $U(x \pm ct)$. This property was first noted by Fisher [13] in the context of population genetics. It was applied to ecological invasions by Skellam [44] in a paper that was also one of those where reaction–diffusion equations were introduced as models for critical patch size. Furthermore, for many equations and some systems, it can be shown that all positive solutions with initial data that are zero outside of some finite interval must propagate (in some sense) with a certain speed c^*. (That idea was introduced by Aronson and Weinberger [2, 3].) These properties are shared by many but not all integro-difference models, see [23, 24]. Traveling waves and other topics related to models for biological invasions are discussed in [11, 16, 28, 37, 43]. A recent survey of invasion theory is given in [17].

For the model

$$\frac{\partial u}{\partial t} = D \frac{\partial^2 u}{\partial x^2} + f(u) \text{ on } \mathbb{R} \times (0, \infty) \tag{3.64}$$

a traveling wavefront that describes an invasion moving from right to left would have the form $u(x,t) = U(x + ct)$, where U is positive and increasing; substituting into (3.64) we find that $U(z)$ should satisfy

$$D \frac{d^2U}{dz^2} - c \frac{dU}{dz} + f(U) = 0. \tag{3.65}$$

For the case of simple linear growth, $f(U) = rU$, solutions of (3.65) will be exponentials. Those that are positive and increasing must have the form $U(z) = U_0 e^{\alpha z}$ for some $\alpha > 0$. Substituting into (3.65) and dividing out U_0 yields $0 = D\alpha^2 - c\alpha + r$, which has solutions $\alpha = (c \pm \sqrt{c^2 - 4rD})/2D$. To have $\alpha > 0$ requires

$$c \geq 2\sqrt{rD}. \tag{3.66}$$

Thus, (3.65) will have wavelike (but unbounded) solutions for $f(u) = ru$ if and only if (3.66) holds. It turns out that the quantity $2\sqrt{rD}$ (or more generally $2\sqrt{Df'(0)}$) is fundamental in the theory of wavefronts for reaction–diffusion models, where $f(u)/u$ is nonincreasing and $f'(0) > 0$. Those include logistic models, among others.

For the general case of (3.65) the existence or nonexistence of wavefronts often can be determined by phase plane analysis. If we let $P = U'$ then (3.65) becomes

$$U' = P$$
$$\tag{3.67}$$
$$P' = -f(U)/D + (c/D)P.$$

Suppose that $f(0) = 0$ and $f(U^*) = 0$ for some $U^* > 0$, so that 0 and U^* are equilibria for (3.64) and (3.65). A wavefront corresponding to an invasion typically would correspond to a solution of (3.65) with $U(z) > 0$, $U(z) \to 0$ as $z \to -\infty$, $U(z) \to U^*$ as $z \to \infty$, $U'(z) \to 0$ as $z \to \pm\infty$, and $U'(z) > 0$. In the phase plane for (3.67) this would correspond to a trajectory lying in the first quadrant that approaches $(0,0)$ as $z \to -\infty$ and $(U^*, 0)$ as $z \to \infty$.

For such trajectories to exist it is necessary that $(0,0)$ be an unstable node or a saddle point with the unstable direction pointing into the first quadrant and that $(U^*, 0)$ be a stable node or saddle point. The linearization of (67) about $(0,0)$ is

$$\begin{pmatrix} U \\ P \end{pmatrix}' = \begin{pmatrix} 0 & 1 \\ -f'(0)/D & c/D \end{pmatrix} \begin{pmatrix} U \\ P \end{pmatrix}. \tag{3.68}$$

The eigenvalues of the matrix in (3.68) are

$$\lambda = \frac{c \pm \sqrt{c^2 - 4Df'(0)}}{2D}. \tag{3.69}$$

It follows that if $f'(0) > 0$ (as in the logistic case) we must have $c \geq 2\sqrt{Df'(0)}$ so that $(0,0)$ will be an unstable node, giving a minimum wave speed as in the linear case. If $f'(0) < 0$, which will be the case for some models will Allee effects, $(0,0)$ will be a saddle point, so to say anything about the wave speed requires deeper analysis. It is reasonable to expect solutions of (64) to approach U^* only if U^* is an asymptotically stable equilibrium of $y' = f(y)$; in that case $f'(U^*) < 0$, so normally $(U^*, 0)$ is a saddle point. That is consistent with the existence of wavefronts but by itself is not sufficient.

The existence of wavefronts can be established by using geometric methods to analyze the phase portrait of (3.67), see [3, 11, 22]. For the case where $f(u) > 0$ for $0 < u < U^*$, as in logistic models, it turns out that the condition $c \geq 2\sqrt{Df^*}$ where

$$f^* = \sup\{f(u)/u : u \in [0, U^*]\} \tag{3.70}$$

is sufficient for the existence of a wavefront of speed c. For the logistic equation itself $f^* = f'(0) = r$ so the condition $c \geq 2\sqrt{rD}$ is necessary and sufficient for the existence of a wavefront with speed c. If $f'(0) > 0$ but $f(u)/u > f'(0)$ for some $u \in (0, U^*)$ (so that there is a weak Allee effect) the minimum wavespeed may be larger than $2\sqrt{Df'(0)}$. When $f'(0) > 0$ there will be wavefronts for all speeds greater than the minimum speed. In the bistable case where $f(0) = f(U^*) = 0$, $f'(0) < 0$, $f'(U^*) < 0$ the situation is more delicate. Suppose that $f(u)$ satisfies those inequalities with $f(u) < 0$ on $(0, a)$, $f(u) > 0$ on (a, U^*) for some $a \in (0, U^*)$, so that (3.64) describes a population with an Allee effect. In that case it turns out that there is a unique speed c for which a wavefront exits, and c may be positive, negative, or zero. To understand how c depends on f in that case, suppose that $u = U(x + ct)$ corresponds to a wavefront for (3.64) with $u(x,t) \to 0$ as $x \to -\infty$ and $u(x,t) \to U^*$ as $x \to \infty$. Then U satisfies (3.65), and $(U(z), U'(z)) \to (0,0)$ as $z \to -\infty$, $(U(z), U'(z)) \to (U^*, 0)$ as $z \to \infty$. If we multiply (3.65) by $U'(z)$ and integrate from $-\infty$ to ∞, then assuming that all of the integrals converge we have

$$0 = \int_{-\infty}^{\infty} DU''(z)U'(z)\,dz - c\int_{-\infty}^{\infty} U'(z)^2\,dz + \int_{-\infty}^{\infty} f(U(z))U'(z)\,dz. \tag{3.71}$$

(The integrals can be shown to converge but we will not consider that here.) If we apply the substitution rule and the conditions at $\pm\infty$ to (3.71) the first integral drops out and the remaining terms can be rearranged to yield

$$c\int_{-\infty}^{\infty} U'(z)^2 \,\mathrm{d}z = \int_0^{U^*} f(U)\,\mathrm{d}U. \qquad (3.72)$$

It follows from (3.72) that the direction of propagation of a wavefront is determined by the sign of the integral of $f(U)$ from 0 to U^*. If the integral is positive then $c > 0$ and the wavefront advances, but if the integral is negative then $c < 0$ and it retreats. If the integral is zero the model typically admits a standing wave. These observations suggest that the presence of Allee effects may be important in determining whether invasions can be stopped. That is indeed the case, see [37].

Systems of reaction–diffusion equations, including models for competition, mutualism, and predator–prey interactions, also admit wavefronts. For models with two or more species the analysis becomes more difficult because the system analogous to (3.67) will typically have a phase space of dimension four or higher. (If one of the species is sedentary it may be three dimensional.) In some cases systems may admit solutions where there is an advancing front followed by periodic oscillations. Also, in the case of systems, it can be difficult to compute the speed of wavelike solutions because of hidden Allee effects. Traveling waves in systems are discussed in [9, 20, 28, 34].

A different class of models that can support wavefronts of constant speed, but also can support accelerating wavefronts, are integrodifference models. Those are discrete time models where a population is assumed to reproduce at each time step and disperse between episodes of reproduction in a way that can be described by an integral kernel. For a population with linear growth rate R in a homogeneous one-dimensional environment an integrodifference model would have the form

$$u(x, t+1) = \int_{-\infty}^{\infty} K(x-y)Ru(y,t)\,\mathrm{d}y \qquad (3.73)$$

for $x \in \mathbb{R}, \ t = 0, 1, 2\ldots.$

If the dispersal process is simple diffusion then $K(x)$ will be a Gaussian of the form shown in (3.42) (with $t = 1$), but a key feature of models such as (3.73) is that $K(x)$ can be adjusted to describe other sorts of dispersal, and can be constructed from empirical data. See [17, 23, 24, 27] for further discussion of these points. If the kernel $K(x)$ is exponentially bounded it will support wavefronts of constant speed. The speed can be calculated in terms of the moment generating function

$$M(s) = \int_{-\infty}^{\infty} K(x)\,\mathrm{e}^{sx}\,\mathrm{d}x. \qquad (3.74)$$

If $K(x)$ is not exponentially bounded then models such as (3.73) may support accelerating waves. To analyze the possible speeds for waves in (3.73) when $K(x)$ is exponentially bounded we would proceed as in the derivation of (3.66) by looking for solutions of (3.73) in the form $u(x, t) = e^{\alpha(x+ct)}$. Some calculations then yield

$$c(\alpha) = \frac{\ln(RM(\alpha))}{\alpha}, \tag{3.75}$$

which typically leads to a minimum wave speed analogous to (3.66). If $K(x)$ is not exponentially bounded (3.73) may have propagating solutions that accelerate rather than moving at a constant rate. These points and other features of models such as (3.73) and its generalizations are discussed in [17, 23, 24].

3.3.2 Propagation and Stability of Solutions

Traveling waves are a special class of solutions to models such as (3.64), but in many cases it is possible to gain information about solutions in general by means of comparison theorems. In models with Allee effects we can say something immediately. Specifically, if $f(u) < 0$ for $0 < u < a$ then any solution to $dy/dt = f(y)$ with $y(0) < a$ will approach zero as $t \to \infty$, and $u(x, t) = y(t)$ will satisfy (3.64), so by the comparison principle (Theorem 2) any positive solution of (3.64) with $u(x, 0) \leq a - \epsilon$ for some $\epsilon > 0$ must approach zero as $t \to \infty$. Thus, propagation is possible only if the initial data are larger than a somewhere. In contrast, if $f'(0) > 0$ then typically all positive solutions whose initial data are zero outside of some half line but positive somewhere will propagate. (If the initial data are positive everywhere the solution might just increase uniformly.) In either case, those solutions that do propagate generally have a characteristic speed of propagation c^*. Such results were obtained by Aronson and Weinberger [3, 4] by means of comparison principles similar to Theorems 2 and 3. We will sketch the general outlines of that approach.

Suppose that $f(u) = r(1 - [u/K])u$. For any solution of (3.64) with initial data satisfying $0 \leq u(x, 0) \leq K$ and $u(x, 0) > 0$ on some interval, the strong maximum principle (Theorem 1, part ii)) implies that $u(x, t) > 0$ for positive t. It is easy to see that if $L > \sqrt{D/r}$ and $\epsilon > 0$ is sufficiently small then $\phi(x) = \epsilon \cos(x/L)$ is a subsolution to the equilibrium problem for (3.64) on $(-\pi L/2, \pi L/2)$, with $\phi(\pm \pi L/2) = 0$. By arguing as in the proof of Theorem 3 it can be seen that the solution to (3.64) with initial data $u(x, 0) = \phi(x)$ on $(-\pi L/2, \pi L/2)$, $u(x, 0) = 0$ otherwise, will increase toward a global equilibrium of (3.64). Since $\epsilon > 0$ is arbitrary, any positive solution of (3.64) is bounded below by such an increasing solution. Any bounded solution of (3.64) will be bounded above by a solution with initial data $u(x, 0) \equiv U_1 > K$ for some U_1; those solutions approach K as $t \to \infty$. It follows that any positive solution of (3.64) is bounded above by solutions approaching K and below by solutions approaching some positive global equilibrium. The equilibria of

(3.64) can be analyzed by observing that if the equilibrium equation is multiplied by du/dx it can be integrated to yield

$$(D/2)\left(\frac{du}{dx}\right)^2 + F(u) = C, \tag{3.76}$$

where $F(u)$ is an antiderivative of $f(u)$. That equation can be analyzed further to show that for $f(u) = r(1 - [u/K])u$ there are no positive global equilibria other than $u \equiv K$, so solutions to (3.64) with nonzero nonnegative initial data all approach K as $t \to \infty$. Since (3.64) supports traveling wavefronts in this case, the convergence generally will not be uniform. In the case of models with Allee effects, analysis of (3.76) can be used to construct subsolutions which can be used to show that solutions to (3.64) whose initial data are large enough over a long enough (but finite) interval will approach the stable positive equilibrium of $dy/dt = f(y)$ as $t \to \infty$.

Once we know that reaction–diffusion models support wavefronts, and that solutions with positive initial data (for logistic-type models) or sufficiently large initial data (for models with Allee effects) propagate, it is natural to ask whether propagating solutions necessarily act like wavefronts. It turns out that they do for models where $f(u)$ is of logistic or bistable (Allee) type, in the sense that solutions with initial data that are zero outside a half line but positive somewhere will propagate with a fixed characteristic speed $c^* > 0$. The speed c^* depends only on $f(u)$ and D. To be precise, suppose that $f(u)$ has $f(0) = 0$, $f(U^*) = 0$ for some $U^* > 0$, and either

$$f'(0) > 0, \ f'(U^*) < 0, \ f(u) > 0 \ \text{on} \ (0, U^*) \tag{3.77}$$

(logistic type) or (for some $a \in (0, U^*)$)

$$f(a) = 0, \ f'(0) < 0, \ f(u) < 0 \ \text{on} \ (0, a),$$

$$f'(a) > 0, \ f(u) > 0 \ \text{on} \ (a, U^*), \ f'(U^*) < 0 \tag{3.78}$$

(Allee or bistable type). Then there exists $c^* > 0$ such that for solutions $u(x, t)$ of (3.64) with initial data $u(x, 0) \geq 0$, $u(x, 0) > 0$ on some interval, $u(x, 0) = 0$ on $(-\infty, x_0)$ for some $x_0 \in \mathbb{R}$, such that $u(x, t) \to U^*$ as $t \to \infty$, we have

$$\lim_{t \to \infty} u(x + ct, t) = 0 \ \text{if} \ c > c^*$$

$$\lim_{t \to \infty} u(x + ct, t) = U^* \ \text{if} \ c < c^*. \tag{3.79}$$

(In the case (3.77) the condition $u(x, t) \to U^*$ as $t \to \infty$ is automatically satisfied, but in (3.78) it requires $u(x, 0)$ to be sufficiently large over a sufficiently long interval.) It turns out that $4Df'(0) \leq (c^*)^2 \leq 4Df^*$ where f^* is defined in (3.70). The meaning of (3.79) is that an observer moving to the left at a speed $c > c^*$ will outrun the propagating solution of (3.64) while if $c < c^*$ the solution will outrun the observer. In that sense the solution must propagate with speed c^*.

The conclusions of (3.79) are obtained by rewriting (3.64) in terms of $z = x + ct$ and t to obtain

$$\frac{\partial u}{\partial t} = D\frac{\partial^2 u}{\partial z^2} - c\frac{\partial u}{\partial z} + f(u). \qquad (3.80)$$

Sub- and supersolutions to (3.65), the equilibrium problem for (3.80), can then be constructed by means of a phase plane analysis of (3.67) similar to what is used to show the existence of wavefronts. For details see [3]. The property of having a fixed propagation speed extends to some reaction–diffusion systems and integrodifference models, see [23, 24, 28].

The stability of wavefronts can be studied by using sub- and supersolutions in (3.80), among other approaches. It turns out that in the case (3.78), where there is a unique speed for wavefronts, solutions of (3.64) with initial data $u(x,0) > a$ as $x \to \infty$, $u(x,0) < a$ as $x \to -\infty$ approach some translated wavefront $U(x + ct + x_1)$ uniformly. For the case (3.77), wavefronts with the minimal wavespeed are stable in a certain sense in terms of their profile but the nature of the convergence is weaker than uniform convergence. For details see [12] or [11] and the references therein.

3.4 Pattern Formation

3.4.1 Types of Patterns

One of the fundamental goals of theoretical biology is to connect the complex patterns observed in the structure of organisms and the distribution of populations to the processes that produce them. Reaction–diffusion models and their spatially discrete analogs describe a class of processes that can lead to the formation and/or persistence of spatial patterns. Specifically, in situations where equilibria for the reaction terms without diffusion are also spatially constant, equilibria for the system with diffusion, as in the case of systems with homogeneous Neumann boundary conditions, adding diffusion can sometimes create instabilities that result in the spontaneous formation of spatial patterns. This was first observed by Turing [49] in 1952, and his name is often used to denote instabilities and patterns arising from that mechanism. A different sort of patterns can arise when the reaction system has multiple stable equilibria, as in the case of models with Allee effects. In that case the corresponding reaction–diffusion models can sometimes support stable solutions that are close to one stable equilibrium of the reaction system on part of their spatial domain but close to another on another part of the domain. For that phenomenon to occur on a bounded spatial domain the domain must be nonconvex, see for example [29, 30]. More recently it has been noted that reaction–diffusion systems can sometimes support solutions that may vary in complex ways in both space and time [32, 39]. Here we will focus on Turing instabilities and patterns for systems on bounded domains with no-flux boundary conditions. More general and detailed treatments are given in [16, 34]. Some treatments of pattern formation that make direct connections to ecological patterns are given in [26, 33, 34, 42].

3.4.2 Diffusion as a Stabilizing Factor

In the case of a single reaction–diffusion equation and for some types of systems Turing instabilities can be ruled out, since diffusion has a stabilizing effect on spatially constant equilibria. (Other sorts of spatial structures are not ruled out in those cases; for example, a single equation with a bistable nonlinearity may support standing waves.)

Consider the single equation

$$\frac{\partial u}{\partial t} = d\nabla^2 u + f(u) \text{ in } \Omega \times (0, \infty)$$

$$\frac{\partial u}{\partial n} = 0 \qquad\qquad \text{on } \partial\Omega \times (0, \infty).$$
(3.81)

If $f(u^*) = 0$ then u^* is an equilibrium of (3.81) and of the corresponding ODE

$$\frac{du}{dt} = f(u).$$
(3.82)

As an equilibrium of (3.82) u^* is stable if $\sigma < 0$ in the linearized eigenvalue problem

$$f'(u^*)v = \sigma v.$$
(3.83)

In (3.83) we have $\sigma = f'(u^*)$ so u^* is stable if $f'(u^*) < 0$ and unstable if $f'(u^*) > 0$. The linearized problem for (3.82) yields

$$d\nabla^2 v + f'(u^*)v = \sigma v \text{ in } \Omega$$

$$\frac{\partial v}{\partial u} = 0 \qquad\qquad \text{on } \partial\Omega,$$

which can be written as

$$d\nabla^2 v = [\sigma - f'(u^*)]v \text{ in } \Omega,$$

$$\frac{\partial v}{\partial n} = 0 \qquad\qquad \text{on } \partial\Omega.$$
(3.84)

From (3.84) it follows that $\dfrac{\sigma - f'(u^*)}{d} = -\lambda_k$ for some eigenvalue λ_k of the problem

$$\nabla^2 v + \lambda v = 0 \text{ in } \Omega$$

$$\frac{\partial v}{\partial \mathbf{n}} = 0 \qquad \text{on } \partial \Omega. \tag{3.85}$$

It is a standard result in the theory of linear PDE that the eigenvalues of (3.85) are $0 = \lambda_1 < \lambda_2 \leq \lambda_3 \ldots$. We can solve for σ to obtain

$$\sigma = f'(u^*) - d\lambda_k, \tag{3.86}$$

so if $f'(u^*) < 0$ then $\sigma < 0$ as well. Thus, diffusion has a stabilizing effect.

For the system

$$\frac{\partial u_i}{\partial t} = d_i \nabla^2 u_i + f_i(\mathbf{u}) \text{ in } \Omega \times (0, \infty)$$

$$\frac{\partial u_i}{\partial \mathbf{n}} = 0 \qquad \text{on } \partial \Omega \times (0, \infty), \quad i = 1, \ldots, N, \tag{3.87}$$

$$i = 1, \ldots, N$$

something similar is true if either $d_1 = d_2 = \cdots = d_N$ or if $\mathbf{f}(\mathbf{u}) = \nabla F(\mathbf{u})$ for some F. The linearized system for (3.87) leads to the eigenvalue problem

$$D\nabla^2 \mathbf{v} + J\mathbf{v} = \sigma \mathbf{v} \text{ in } \Omega,$$

$$\frac{\partial v_i}{\partial \mathbf{n}} = 0 \qquad \text{on } \partial \Omega, \quad i = 1, \ldots, N, \tag{3.88}$$

where D is a diagonal matrix with diagonal entries $d_1, \ldots, d_N$ and J is the Jacobian of $\mathbf{f}(\mathbf{u})$ evaluated at the equilibrium of interest $\mathbf{u}^*$. The equilibrium $\mathbf{u}^*$ will be asymptotically stable in the system of ordinary differential equations corresponding to (3.88) without diffusion if the real parts of the eigenvalues of J are negative. Let S be a matrix such that $S^{-1}JS = M$ with M in Jordan canonical form. If $d_1 = d_2 = \cdots = d_N = d$ so $D = dI$ then $S^{-1}DS = dI$. Applying the similarity transformation to (3.88) in that case yields

$$dI\nabla^2 \mathbf{v} + M\mathbf{v} = \sigma \mathbf{v}. \tag{3.89}$$

If the eigenvalues of J are $\mu_1, \mu_2, \ldots, \mu_k$ then (3.89) decouples by Jordan blocks, but the decoupled equations include for each μ_j

$$d\nabla^2 v_i + \mu_j v_i = \sigma v_i \text{ for some } i. \tag{3.90}$$

The eigenvalues of (3.90) have the form $\sigma = \mu_j - d\lambda_\ell$ where $\lambda_\ell \geq 0$ is an eigenvalue of (3.85). Thus, the real parts of eigenvalues of (3.88) are less than those of the eigenvalues μ_j of J, so if $Re\mu_j < 0$ for all eigenvalues of J then the same is true for eigenvalues of (3.88), so diffusion cannot destabilize the system in that case.

The analysis of the case where the diffusion coefficients of (3.87) are distinct but $\mathbf{f}(\mathbf{u})$ is a gradient so that J is symmetric follows the same line of argument used to show that the eigenvalues of a formally self-adjoint differential operator are real. The essential idea is to multiply (3.88) by the complex conjugate of $\mathbf{v}$, integrate over Ω, and use the symmetry of J, (3.88) and the complex conjugate of (3.88) to manipulate the resulting integrals. Recall that the eigenvalues of a symmetric matrix are real. If μ_1 is the largest eigenvalue of J, the argument described above shows that σ is real and $\sigma \leq \mu_1$. It follows that in that case that if $\mu_1 < 0$ so that $\mathbf{u}^*$ is an asymptotically stable equilibrium of the system without diffusion then $\mathbf{u}^*$ cannot be destabilized by diffusion. Thus, Turing instabilities require that $d_i \neq d_j$ for some i and j and that J is not symmetric. (As noted previously, those conditions are not necessary for the existence of stable nonconstant solutions, but if they are not satisfied such solutions must arise from some other mechanism.)

3.4.3 Diffusion as a Destabilizing Mechanism

From the preceding discussion it follows that the simplest case of (3.88) for which we might obtain diffusion-driven instability can be written as

$$
\begin{aligned}
\nabla^2 v_1 + a v_1 + b v_2 &= \sigma v_1, \\
\delta \nabla^2 v_2 + c v_1 + d v_2 &= \sigma v_2 \text{ in } \Omega, \\
\frac{\partial v_1}{\partial \mathbf{n}} = \frac{\partial v_2}{\partial \mathbf{n}} &= 0 \qquad \text{on } \partial \Omega.
\end{aligned} \tag{3.91}
$$

The corresponding eigenvalue problem for the system without diffusion is simply

$$
J \mathbf{v} = \begin{pmatrix} a & b \\ c & d \end{pmatrix} \mathbf{v} = \sigma \mathbf{v}. \tag{3.92}
$$

The eigenvalues of J are

$$
\mu = \frac{(a+d) \pm \sqrt{(a+d)^2 - 4(ad - bc)}}{2}.
$$

These will have negative real parts if

$$
a + d < 0, \quad ad - bc > 0. \tag{3.93}
$$

If (3.93) holds then $(0,0)$ is stable for the system of ordinary differential equations

$$
\begin{aligned}
\frac{du}{dt} &= au + bv, \\
\frac{dv}{dt} &= cu + dv.
\end{aligned} \tag{3.94}
$$

Suppose (3.93) holds. We can find eigenvalues for (3.94) by using eigenfunctions of ∇^2. Let $\phi(x)$ be an eigenfunction for (3.85) with eigenvalue λ. Let $\mathbf{v} = \begin{pmatrix} p \\ q \end{pmatrix} \phi(x)$ for some constants p and q. Substituting $\mathbf{v}$ into (3.91) yields

$$\begin{pmatrix} a - \lambda & b \\ c & d - \delta\lambda \end{pmatrix} \begin{pmatrix} p \\ q \end{pmatrix} = \sigma \begin{pmatrix} p \\ q \end{pmatrix} \tag{3.95}$$

so that σ is an eigenvalue of

$$\begin{pmatrix} a - \lambda & b \\ c & d - \delta\lambda \end{pmatrix}. \tag{3.96}$$

It follows that:

$$\sigma = \frac{(a + d) - (\delta + 1)\lambda \pm \sqrt{[(a + d) - (\delta + 1)\lambda]^2 - 4[(a - \lambda)(d - \delta\lambda) - bc]}}{2}.$$

$$\tag{3.97}$$

By (3.93) and the nonnegativity of eigenvalues of (3.85) we have

$$a + d - (\delta + 1)\lambda < 0.$$

Hence, to obtain diffusion-driven instability we must have

$$(a - \lambda)(d - \delta\lambda) - bc < 0. \tag{3.98}$$

Note that the eigenvalues of (3.85) can have any nonnegative value. If $\Omega = (0, \ell)$ then $\lambda_k = \dfrac{\pi^2 k^2}{\ell^2}$ is an eigenvalue of (3.85) for $k = 0, 1, 2, \ldots$. Thus, we will have established the possibility of diffusion-driven instability if there are values of λ for which (3.98) holds, given that (3.93) is satisfied. Whether or not diffusion-driven instability will actually occur depends on whether or not there are eigenvalues of (3.85) satisfying (3.98), which in turn depends on Ω. (We will return to this point later.)

Note that (3.93) implies that (3.98) cannot hold for $\lambda = 0$, and that for large positive values of λ the left side of (3.98) is positive so (3.98) cannot hold. The case where (3.98) will hold for some positive values of λ is when the quadratic in λ on the left has positive real roots; then (3.98) holds for λ between those roots. The quadratic is $\delta\lambda^2 - (\delta a + d)\lambda + ad - bc$, so the roots are

$$\lambda = \frac{(\delta a + d) \pm \sqrt{(\delta a + d)^2 - 4(ad - bc)\delta}}{2\delta}. \tag{3.99}$$

By (3.93), $ad - bc > 0$ and $a + d < 0$; hence, (3.99) can have positive roots if and only if $\delta a + d > 0$, which is consistent with $a + d < 0$ only if a and d have opposite signs. If a and d have opposite signs then since we assumed $ad - bc > 0$ in (3.93), b and c must have opposite signs as well. Thus, for diffusion-driven instability of the equilibrium $(0, 0)$ of the system

$$\frac{\partial u_1}{\partial t} = \nabla^2 u_1 + a u_1 + b u_2$$

$$\frac{\partial u_2}{\partial t} = \delta \nabla^2 u_2 + c u_1 + d u_2 \ \text{ in } \ \Omega \times (0, \infty), \tag{3.100}$$

$$\frac{\partial u_i}{\partial \mathbf{n}} = 0 \qquad\qquad \text{on } \partial\Omega \times (0, \infty), \ i = 1, 2,$$

we must have

(i) $a + d < 0$, $ad - bc > 0$, so that $(0,0)$ is stable for the ODE system (3.94);
(ii) $\delta a + d > 0$, so that (3.99) has positive roots $\lambda_\pm$.
(iii) Ω is such that ∇^2 with no-flux boundary conditions has one or more eigenvalues $\lambda \in (\lambda_-, \lambda_+)$.

For the first two of these conditions to hold, a and d must have opposite signs, and b and c must have opposite signs. The possible patterns are

$$\begin{pmatrix} + & - \\ + & - \end{pmatrix}, \begin{pmatrix} - & + \\ - & + \end{pmatrix}, \begin{pmatrix} + & + \\ - & - \end{pmatrix}, \begin{pmatrix} - & - \\ + & + \end{pmatrix}.$$

The coefficients b and c describe how each component affects the other near equilibrium. If $b > 0$ and $c < 0$ then the second component tends to make the first increase and the first tends to make the second decrease. In applied settings the first component would be said to be an inhibitor and the second an activator. (If the signs are reversed the roles are reversed.) If $a > 0$ then the first component contributes to its own growth; in chemical applications this is sometimes called autocatalysis, and in ecological models it is a version of an Allee effect. If $a > 0$ then we must have $d < 0$, which indicates that the second component contributes to its own decline; this is something like a logistic effect or intraspecific interference.

The requirement that $\delta a + d > 0$ means that if $a > 0$ we must have $\delta > 1$, so that the second component must diffuse more rapidly than the first; if $a < 0$ then $\delta < 1$ is needed so the first component diffuses faster than the second. Thus, the autocatalytic component must diffuse more slowly than the self-limiting component for diffusion-driven instability to occur. In many applications the autocatalytic component stimulates or activates the self-limiting component while the self-limiting component inhibits the autocatalytic component. In those situations the inhibitor must diffuse more rapidly than the activator.

Example (from ecology). We will look at some ecological models in one space dimension, so $\Omega = (0, \ell)$. The first observation is negative. Consider the Lotka–Volterra model

$$\frac{du}{dt} = (A + Bu + Cv)u$$

$$\frac{dv}{dt} = (D + Eu + Fv)v.$$

(3.101)

Normally $B, F \leq 0$ (either one or both species display logistic self-limitation) but the signs of the other coefficients depend on whether the model describes mutualism, competition, or a predator–prey interaction. If (u^*, v^*) is a positive equilibrium of (3.101) then $A + Bu^* + Cv^* = D + Eu^* + Fv^* = 0$ and

$$J = \begin{pmatrix} Bu^* & Cu^* \\ Ev^* & Fv^* \end{pmatrix}.$$

Since B and F are both nonpositive, diffusion cannot destabilize a Lotka–Volterra system. However, if we allow an Allee effect then diffusion-driven instability is possible. The following system could arise as a model for a predator–prey system, where u is the prey and v the predator, u has an Allee effect, and v has a logistic self-limitation effect:

$$\frac{du}{dt} = [-1 + 9u - 4u^2 - 4v]u = -u + 9u^2 - 4u^3 - 4uv$$

$$\frac{dv}{dt} = [3u - 1 - 2v]v = 3uv - v - 2v^2.$$

(3.102)

The sort of assumptions made in (3.102) are fairly typical of those which have been made in models for predator–prey systems with pattern formation, for example [30, 42]. Another sort of assumption that can produce pattern formation is mutual interference by predators [1].

The system (3.102) has equilibrium $(1, 1)$. Linearizing at $(1, 1)$ yields

$$J = \begin{pmatrix} -1 + 18u - 12u^2 - 4v & -4u \\ 3v & 3u - 1 - 4v \end{pmatrix} \Big|_{(1,1)}$$

$$= \begin{pmatrix} 1 & -4 \\ 3 & -2 \end{pmatrix}.$$

We have $a + d = -1$, $ad - bc = 10$, so $(1, 1)$ is stable for (3.102). We have $\delta a + d = \delta - 2 > 0$ if $\delta > 2$, so in that case the system

$$\frac{\partial u}{\partial t} = \frac{\partial^2 u}{\partial x^2} + [-1 + 9u - 4u^2 - 4v]u$$

$$\frac{\partial v}{\partial t} = \delta \frac{\partial^2 v}{\partial x^2} + [3u - 1 - 2v]v \qquad \text{on } (0, \ell) \times (0, \infty)$$

(3.103)

$$\frac{\partial u}{\partial x} = \frac{\partial v}{\partial x} = 0 \qquad \text{at } x = 0, \ell$$

may have diffusion-driven instabilities. The expression in (3.99) in this case is

$$\lambda_\pm = \frac{(\delta - 2) \pm \sqrt{(\delta - 2)^2 - 40\delta}}{2\delta}.$$

(3.104)

Thus, the system has the possibility of Turing instability if $(\delta - d)^2 - 40\delta > 0$, which will be true for δ sufficiently large. The eigenvalues for (3.85) in this case are $\lambda_k = \pi^2 k^2 / \ell^2$, $k = 0, 1, \ldots$. If $\pi^2/\ell^2 > \lambda_+$ then there are no eigenvalues of (3.85) in the interval (λ_-, λ_+), and so the domain $\Omega = (0, \ell)$ is too small to support Turing instabilities. In general there is a critical patch size needed to support the formation of Turing patterns, just as there is a critical patch size needed to support a population when there is loss of population across the patch boundary. In both cases the critical patch size is related to the eigenvalues of the operator ∇^2. As ℓ increases, the eigenvalues λ_k decrease so that they first enter (λ_-, λ_+) at λ_+ then eventually exit through λ_-. For any given value of ℓ there will only be finitely many values of $\lambda_k \in (\lambda_-, \lambda_+)$. The initial patterns that arise from the diffusion-driven instability will resemble the eigenfunctions $\phi_k(x) = \cos(\pi k x/\ell)$ corresponding to the eigenvalues $\lambda_k \in (\lambda_-, \lambda_+)$, so larger domains will support more complex patterns. In the one-dimensional case the eigenfunction corresponding to λ_k changes sign k times, so k is sometimes called the wavenumber. If we substitute $\lambda_k = \pi^2 k^2/\ell^2$ into (3.94) we obtain a formula $\sigma = \sigma(k)$, which relates stability to wavenumber. This is sometimes called a dispersion relation. Dispersion relations can be used to study the possibility of pattern formation when the underlying spatial domain is $(-\infty, \infty)$. An approach that is sometimes used in that context is to look for solutions of the linearized model in the form $\mathbf{u} = \mathbf{u}_0\, e^{\sigma t \pm i \lambda x}$ and thus derive a dispersion relation. For additional discussion pattern formation see [16, 34].

3.5 Additional Topics and Resources in Reaction–Diffusion Theory

The discussion presented here treats only the basic aspects of what is now the "classical" theory of reaction–diffusion equations and systems. All of the topics we have discussed are treated in greater depth and/or breadth in other sources, and there are more modern aspects of the theory that we have not addressed in any detail. Specifically, reaction–diffusion equations and systems can often be cast as semidynamical systems and analyzed by methods from the theory of dynamical systems and nonlinear functional analysis. Here we will give a brief, informal discussion of some modern approaches and cite some references for further study of those and the "classical" theory.

The formulation of reaction–diffusion models as semidynamical systems is typically accomplished by means of the theory of semigroups of operators. Semigroup theory is presented in [15, 40]. The application of that theory to the formulation of reaction–diffusion models as semidynamical systems is given in [18]. Although reaction–diffusion models have infinite dimensional state spaces, the regularity theory of parabolic partial differential equations typically implies that the trajectories of bounded solutions are precompact. Because of that feature much of the theory of finite dimensional dynamical systems extends to reaction–diffusion systems. One aspect of that theory that is especially relevant to ecological models is the notion of permanence or uniform persistence. Roughly speaking, a system is permanent if it is cast on a state space with a positive cone which has a nonempty interior, the positive cone is invariant (at least as time goes forward), and there is an attractor which lies inside the positive cone, is bounded away from the boundary of the positive cone, and is globally attracting among positive solutions. A system that is permanent need not have a stable positive equilibrium, and may have complex dynamics, but it predicts that if all its components are initially positive then eventually they will all have positive lower bounds in an appropriate metric. Ecologically, a prediction of permanence means that no species originally present will become extinct, and that the density of each species will eventually exceed some lower bound independent of its initial density. Thus, permanence extends the idea of stable coexistence to situations where the system does not necessarily have an equilibrium that is globally attracting among positive solutions. Applications of permanence to reaction–diffusion models in ecology are treated in [6]. Another approach that is widely used to study dynamical systems and extends to reaction–diffusion models is bifurcation theory. Since the state spaces of reaction–diffusion models are infinite dimensional, the framework of nonlinear functional analysis is needed to set up the bifurcation theory. Discussions of bifurcation theory for reaction–diffusion models are given in [6, 46]. Finally, results based on comparison principles can be combined with the dynamical systems approach to study reaction–diffusion models from the viewpoint of monotone or order preserving dynamical systems. Such approaches are described in [19, 45].

There are a number of works that treat aspects of reaction–diffusion theory described in this tutorial in greater depth. The formulation and interpretation of reaction–diffusion models in ecology is treated in [6, 34, 35]. Comparison principles, sub- and supersolutions, and other results based on maximum principles are treated in [6, 25, 38, 46]. Traveling waves and pattern formation are discussed in [16, 34]. Connections with dynamical systems theory are described in [6, 18, 19, 45]. This list of references is not intended to be comprehensive, but it should give the interested reader a starting point for the deeper study of reaction–diffusion models.

References

[1]. D. Alonso, F. Bartumeus, and J. Catalan. Mutual interference between predators can give rise to Turing spatial patterns. Ecology, **83**: 28-34, 2002.

[2]. D.A. Andow, P. Kareiva, S.A. Levin, and A. Okubo. Spread of invading organisms. Landscape Ecology, **4**: 177-188, 1990.

[3]. D.G. Aronson and H.F. Weinberger. Nonlinear diffusion in population genetics, combustion, and nerve impulse propagation. In Partial Differential Equations and Related Topics, Lecture Notes in Mathematics 446, pages 5-49. Springer, Berlin, 1975.

[4]. D.G. Aronson and H.F. Weinberger. Multidimensional nonlinear diffusions arising in population genetics. Advances in Math., **30**: 33-76, 1978.

[5]. R.S. Cantrell and C. Cosner. Practical persistence in ecological models via comparison methods. Proc. Royal Soc. Edinburgh, **126A**: 247-272, 1996.

[6]. R.S. Cantrell and C. Cosner. Spatial Ecology via Reaction-Diffusion Equations. Wiley, Chichester, England, 2003.

[7]. C. Cosner. Variability, vagueness, and comparison methods for ecological models. Bull. Math. Biol., **58**: 207-246, 1996.

[8]. C. Cosner. Comparison principles for sytems that embed in cooperative systems, with applications to Lotka-Volterra models. Dyn. Disc., Cont., and Impl. systems **3**: 283-303, 1997.

[9]. S. Dunbar. Traveling waves in diffusive predator-prey equations: periodic orbits and point-to-periodic heteroclinic orbits. SIAM J. Appl. Math., **46**: 1057-1078, 1986.

[10]. R. Durrett and S.A. Levin. The importance of being discrete (and spatial). Theor. Pop. Biol., **46**: 363-394, 1994.

[11]. P. Fife. Mathematical Aspects of Reacting and Diffusing Systems, Lecture Notes in Biomathematics **28**. Springer, Berlin, 1979.

[12]. P. Fife and J.B. McLeod. The approach of solutions of nonlinear diffusion equations to traveling front solutions. Arch. Rat. Mech. Anal., **65**: 335-361, 1977.

[13]. R.A. Fisher. The wave of advance and advantageous genes. Ann. Eugenics, **7**: 353-369, 1937.

[14]. A. Friedman. Partial Differential Equations of Parabolic Type. Prentice Hall, Englewood Cliffs, New Jersey, 1964.

[15]. A. Friedman. Partial Differential Equations. Krieger, New York, 1976.

[16]. P. Grindrod. The Theory and Applications of Reaction-Diffusion Equations: Patterns and Waves, 2nd edition. Oxford University Press, Oxford, 1996.

[17]. A. Hastings, K. Cuddington, K.F. Davies, C.J. Dugaw, E. Elmenderf, A. Freestone, S. Harrison, M. Holland, J. Lambrinos, U. Malvadkar, B.A. Melbourne, K. Moore, C. Taylor, and D. Thompson. The spatial spread of invasions: new developments in theory and evidence. Ecology Letters, **8**: 91-101, 2005.

[18]. D. Henry. Geometric Theory of Semilinear Parabolic Equations, Lecture Notes in Mathematics **840**, Springer, Berlin, 1981.

[19]. P. Hess. Periodic-Parabolic Boundary Value Problems and Positivity, Pitman Research Notes in Mathematics **247**, Longman, Harlow, Essex, UK, 1991.

[20]. Y. Hosono. The minimal speed of traveling fronts for a diffusive Lotka-Volterra competition model. Bull. Math. Biol., **60**: 435-448, 1998.

[21]. H. Kierstead and L.B. Slobodkin. The size of water masses containing plankton bloom. J. Marine Research, **12**: 141-147, 1953.

[22]. M. Kot. Elements of Mathematical Ecology. Cambridge University Press, Cambridge, U.K.

[23]. M. Kot and W.M. Schaeffer. Discrete-time growth-dispersal models. Math. Biosci., **80**: 109-136, 1986.

[24]. M. Kot, M.A. Lewis, and P. van den Driessche. Dispersal data and the spread of invading organisms. Ecology, **77**: 2027-2042, 1996.

[25]. A. Leung. Systems of Nonlinear Partial Differential Equations. Kluwer Academic, Boston, 1989.

[26]. S.A. Levin and L.A. Segel. Pattern generation in space and aspect. SIAM Review, **27**: 45-67, 1985.

[27]. M.A. Lewis. Variability, patchiness, and jump dispersal in the spread of an invading population. In D. Tilman and P. Kareiva, editors, Spatial Ecology, pages 46-69. Princeton University Press, Princeton, New Jersey, 1997.

[28]. M.A. Lewis, B.T. Li, and H.E. Weinberger. Spreading speed and linear determinacy for two-species competition models. J. Math. Biol., **45**: 219-233, 2002.

[29]. H. Matano. Asymptotic behavior and stability of solutions of semilinear diffusion equations. Publ. Res. Inst. Math. Sci. Kyoto, **15**: 401-454, 1979.

[30]. H. Matano and M. Mimura. Pattern formation in competition-diffusion systems in non convex domains. Publ. Res. Inst. Math. Sci. Kyoto, **19**: 1049-1079, 1983.

[31]. R.C. McOwen. Partial Differential Equations: Methods and Applications, 2nd edition Pearson Education, Upper Saddle River, New Jersey, 2003.

[32]. A.B. Medvinski, S.V. Petrovskii, I.A. Tikhonova, H. Malchow, and B.-L. Li. Spatiotemporal complexity of plankton and fish dynamics. SIAM Review, **44**: 311-370, 2002.

[33]. M. Mimura and J.D. Murray. On a diffusive predator-prey model that exhibits patchiness. J. Theor. Biol., **75**: 249-262, 1978.

[34]. J.D. Murray. Mathematical Biology II: Spatial Models and Biomedical Applications. Springer, Berlin, 2003.

[35]. A. Okubo and S.A. Levin. Diffusion and Ecological Problems, 2nd edition. Springer, Berlin, 2001.

[36]. A. Okubo, P.K. Maini, M.H. Williamson, and J.D. Murray. On the spatial spread of the grey squirrel in Britain. Proc. Royal Soc. London, **B238**: 113-125, 1989.

[37]. M.R. Owen and M.A. Lewis. How predation can slow, stop, or reverse a prey invasion. Bull. Math. Biol., **63**: 655-684, 2001.

[38]. C.V. Pao. Nonlinear Parabolic and Elliptic Equations. Plenum Press, New York, 1992.

[39]. M. Pascual. Diffusion-induced chaos in a spatial predator-prey system. Proc. Royal Soc. London, **B251**: 1-7, 1993.

[40]. A. Pazy. Semigroups of Linear Operators and Applications to Partial Differential Equations. Springer, Berlin, 1983.

[41]. M.H. Protter and H.F. Weinberger, Maximum Principles in Differential Equations. Prentice-Hall, Englewood Cliffs, New Jersey, 1967.

[42]. L.A. Segel and J.L. Jackson. Dissipative Structure: an explanation and an ecological example. J. Theor. Biol. **37**: 545-559, 1972.

[43]. N. Shigesada and K. Kawasaki. Biological Invasions: Theory and Practice. Oxford University Press, Oxford, 1997.

[44]. J.G. Skellam. Random dispersal in theoretical populations. Biometrika, **38**: 196-218, 1951.

[45]. H.L. Smith. Monotone Dynamical Systems. Mathematical Surveys and Monographs **41**. American Mathematical Society, Providence, Rhode Island.

[46]. J. Smoller. Shock Waves and Reaction-Diffusion Equations. Springer, Berlin, 1982.

[47]. W.A. Strauss. Partial Differential Equations. Wiley, New York, 1992.

[48]. P. Turchin. Quantitative Analysis of Movement. Sinauer, Sunderland, Massachusetts, 1998.

[49]. A.M. Turing. The chemical basis for morphogenesis. Phil. Trans. Royal Soc. London **B237**: 37-72, 1952.

[50]. P. Yodzis. Introduction to Theoretical Ecology. Harper and Row, New York, 1989.

4

The Dynamics of Migration–Selection Models

T. Nagylaki[1]* and Y. Lou[2]

[1] Department of Ecology and Evolution, The University of Chicago, 1101 East 57th Street, Chicago, IL 60637, USA
[2] Department of Mathematics, The Ohio State University, Columbus, OH 43210, USA
email: choman@uchicago.edu

Summary. The evolution of the gene frequencies at a single multiallelic locus under the joint action of migration and selection is reviewed. The three models treated are in (1) discrete space and discrete time; (2) discrete space and continuous time; and (3) continuous space and continuous time. These models yield, respectively, a system of (1) nonlinear, first-order difference equations; (2) nonlinear, first-order differential equations; and (3) semilinear, parabolic partial differential equations. Among the questions discussed are the loss of a specified allele, the maintenance of a specified allele or every allele, the existence and stability of completely polymorphic equilibria, the weak- and strong-migration limits, and uniform (i.e., location-independent) selection. Many examples and unsolved problems are discussed.

4.1 Introduction

Population genetics concerns the genetic structure and evolution of populations. It has important interfaces with ecology, demography, epidemiology, phylogeny, genomics, and molecular evolution. Its principal applications are to human and medical genetics and to animal and plant breeding.

The genetic composition of a population is usually described by genotypic or allelic frequencies in deterministic models and by the probability distribution of these frequencies (or functionals of such distributions) in stochastic models. These variables may depend on space and time. The mathematical models for their dynamics can be derived rigorously from the basic genetic principles of Mendelism and recombination and the well understood form of a few evolutionary forces, viz., the mating system, selection, mutation, migration, and random genetic drift.

The major theme of theoretical population genetics is the investigation of the amount and pattern of genetic variability under sundry combinations of evolutionary forces. The results are essential for the motivation, design, and

---

* Corresponding author. Tel.: 773-702-1079; fax: 773-702-9740

interpretation of experiments and field studies and for the intelligent numerical analysis of less tractable systems. Comparison of theoretical results with observations often yields information about particular natural populations, and many such comparisons delineate the relative importance and domain of action of the evolutionary forces. In addition to its intrinsic interest and direct influence on empirical research, theoretical work suggests and assists less mathematical, more immediately data-oriented investigations.

Since the genetic mechanisms and the action of the evolutionary forces on each individual are probabilistic, population-genetic models are intrinsically stochastic. However, if the population number or density is large and the parameters in the evolutionary forces are constant or vary deterministically, then for most purposes deterministic models are sufficiently accurate. Nonetheless, even in this case, keeping the underlying stochastic process in mind is often helpful and illuminating.

Since most natural populations are distributed in space and mate at random only locally, it is important to study the amount and pattern of genetic variation in geographically structured populations and to inquire under what conditions genetic variation is independent of geographical structure. In this paper, we examine evolution under the joint action of migration and viability selection; unless otherwise specified, both forces may depend on position but not on time. We ask whether and how spatial variation in the genotypic fitnesses is reflected in spatial variation in the gene frequencies. We investigate also when uniform (i.e., location-independent) fitnesses lead to uniform gene frequencies.

Under the usual assumption that selection is much stronger than mutation, the latter affects the dynamics of very rare alleles only. Therefore, we neglect it. We posit that subpopulation numbers or population densities are sufficiently large to justify neglecting random genetic drift.

The monoecious, diploid population is locally panmictic. Adults migrate independently of genotype. We focus on the evolution of the gene frequencies at a single multiallelic locus, and also present some results for two alleles. The multilocus problem is even more difficult, and most of the literature is intuitive and numerical (Barton and Shpak, 2000, and refs. therein).

Our aim here is to summarize, integrate, and illustrate the results in Nagylaki and Lou (2001, 2006b, 2007a,b) and Lou and Nagylaki (2002, 2004, 2006). For brevity, we refer to these papers by initials and year in the current decade, e.g., NL6b. Here, we present the essence of our assumptions and results; we give specific references to longer, more precise statements. We shall also mention numerous unsolved problems.

For a concise description of the required genetic background and an introduction to theoretical population genetics, some readers may find it helpful to consult Nagylaki (1992).

It is best to develop the theory from the fundamental model with finitely many discrete colonies (or demes) and discrete, nonoverlapping generations, hereafter abbreviated as DD. This approach has three advantages.

First, extension to separate sexes, X-linked loci, and plants with pollen and seed dispersion and the incorporation of random genetic drift can be accomplished straightforwardly (Nagylaki, 1978a,b, 1979, 1996, 1997; Moody, 1979; Nagylaki and Lucier, 1980). Second, only the DD model has a relatively simple, yet illuminating, special case, the Levene (1953) model. Third, the three continuous approximations discusssed below can be derived rigorously from the DD model.

If we retain discrete, nonoverlapping generations, place the colonies at the points of a lattice in d dimensions, and let the lattice spacing tend to 0, the difference equations of the DD model converge to a system of integro-difference equations. Although this continuous-space, discrete-time (CD) model has many applications to continuously distributed populations, it has received relatively little attention (Lui, 1986, and refs. therein); we do not treat it here.

If migration and selection are comparable and both weak, with appropriate time scaling the difference equations of the DD model converge to a system of nonlinear differential equations. In some ways, this discrete-space, continuous-time (DC) model is more tractable than our basic DD scheme. For some organisms without a breeding season, the DC model may be a more accurate approximation.

Suppose now that the colonies are at the points of a lattice in d dimensions, migration and selection are both weak, and migration satisfies the standard assumptions for a diffusion process. If time and the lattice spacing are scaled suitably, the DD difference equations converge to a system of semilinear, parabolic partial differential equations. Many results that have not been proved for discrete space can be established in this continuous-space, continuous-time (CC) diffusion model. For some spatially continuous populations without a breeding season, the CC model may be the most accurate.

Thus, each model has advantages and provides insight. We examine the DD, DC, and CC models in Sects. 4.2, 4.3, and 4.4, respectively. In each section, the results, examples, and unsolved problems follow the formulation of the model. In the Appendix, we list and briefly define the symbols used in this paper.

4.2 Discrete Space and Discrete Time

In Sects. 4.2.1, 4.2.2, 4.2.3, and 4.2.4 we present our formulation of DD model, general results, examples, and our treatment of the Levene model, respectively.

4.2.1 Formulation

We follow the brief formulation of the DD model in NL7a. There is more explanation in NL1; for a much more detailed exposition, consult Nagylaki (1992, Sect. 6.2). Karlin (1978, 1984), Lyubich (1992, Sects. 9.1–9.4), Nagylaki (1992, Sects. 4.1–4.3), and Bürger (2000, Sect. I.9) review the theory of selection at a single locus in a panmictic population; see also Nagylaki and Lou (2006a).

Generations are discrete and nonoverlapping. The monoecious, diploid population is subdivided into K panmictic colonies that exchange adult migrants independently of genotype. Selection acts solely through viability differences: we assume that all fertilities are the same. We neglect mutation and random genetic drift.

We consider a single locus with J alleles A_i. Designating the sets of alleles and demes by $\mathcal{J}$ and $\mathcal{K}$, respectively, we have

$$i \in \mathcal{J} = \{1, 2, \ldots, J\}, \quad k \in \mathcal{K} = \{1, 2, \ldots, K\}. \tag{4.1a}$$

It will be useful to define for every $i \in \mathcal{J}$ and every $k \in \mathcal{K}$

$$\mathcal{J}_i = \{j \in \mathcal{J} : j \neq i\}, \quad \mathcal{K}_k = \{\ell \in \mathcal{K} : \ell \neq k\}. \tag{4.1b}$$

Let $p_{i,k}(t)$ represent the frequency of A_i in zygotes in deme k in generation t $(= 0, 1, 2, \ldots)$. Thus, for every $k \in \mathcal{K}$, the nonnegative variables $p_{i,k}$ satisfy

$$\sum_i p_{i,k} = 1. \tag{4.2}$$

We shall consistently use the subscripts h, i, and j for alleles and k, ℓ, and n for demes.

We denote the simplex by

$$\Delta_J = \left\{ z \in \mathcal{R}^J : z_i \geq 0 \; \forall i \in \mathcal{J}, \; \sum_{j=1}^{J} z_j = 1 \right\}, \tag{4.3}$$

write $\Delta_J^K = (\Delta_J)^K$, and define the column vectors

$$p_i = (p_{i,1}, \ldots, p_{i,K})^{\mathrm{T}} \in \mathcal{R}^K, \tag{4.4a}$$

$$p^{(k)} = (p_{1,k}, \ldots, p_{J,k})^{\mathrm{T}} \in \Delta_J, \tag{4.4b}$$

$$p = (p^{(1)}; \ldots; p^{(K)})^{\mathrm{T}} \in \Delta_J^K. \tag{4.4c}$$

Clearly, p_i, $p^{(k)}$, and p signify the frequencies of A_i in each deme, the gene frequencies in deme k, and all the gene frequencies, respectively.

If the constant $w_{ij,k}$ denotes the viability of an $A_i A_j$ individual in deme k, then the mean viabilities of individuals that carry A_i and of all individuals in deme k are, respectively,

$$w_{i,k}(p^{(k)}) = \sum_j w_{ij,k} p_{j,k}, \quad \bar{w}_k(p^{(k)}) = \sum_{i,j} w_{ij,k} p_{i,k} p_{j,k}. \tag{4.5}$$

Of course, $w_{ij,k} = w_{ji,k}$ for every $i, j \in \mathcal{J}$ and every $k \in \mathcal{K}$. Let $m_{k\ell}$ designate the probability that an individual in deme k after migration came from deme ℓ. Then the gene frequencies in the next generation are given by

$$p'_{i,k} = \sum_\ell m_{k\ell} p^*_{i,\ell}, \tag{4.6a}$$

where

$$p^*_{i,k} = p_{i,k} w_{i,k} / \bar{w}_k \tag{4.6b}$$

represents the frequency of A_i in adults in deme k. Of course, the $K \times K$ backward migration matrix $M = (m_{k\ell})$ is stochastic: $m_{k\ell} \geq 0$ for every $k, \ell \in \mathcal{K}$ and

$$\sum_{\ell} m_{k\ell} = 1 \tag{4.7}$$

for every $k \in \mathcal{K}$.

The retrospective distribution M reflects spatial variation not only of the migration rates, but also of the subpopulation numbers. To separate these two sources of variation, we introduce the prospective distribution $\tilde{M}$.

Let c_k and c^*_k signify the proportion of individuals in deme k before and after selection, respectively. Therefore, we have

$$c = (c_1, \ldots, c_K)^{\mathrm{T}} \in \mathrm{int}\Delta_K, \quad c^* = (c^*_1, \ldots, c^*_K)^{\mathrm{T}} \in \mathrm{int}\Delta_K. \tag{4.8}$$

Let the constant $\tilde{m}_{k\ell}$ denote the probability that an individual in deme k migrates to deme ℓ. The forward migration matrix $\tilde{M} = (\tilde{m}_{k\ell})$ is also stochastic, and for every $k, \ell \in \mathcal{K}$ we have

$$m_{k\ell} = c^*_\ell \tilde{m}_{\ell k} \bigg/ \sum_{n} c^*_n \tilde{m}_{nk}. \tag{4.9}$$

Most analyses have been devoted to two models of adult (i.e., post-selection) migration: soft and hard selection. The former means that during selection, the population is regulated within each deme so that $c^* = c$. In this case, we see from (4.9) that M is constant, and therefore our basic DD model (4.6) is complete.

Hard selection means that it is the total population number that is controlled, in which case

$$c^*_k = c_k \bar{w}_k \bigg/ \sum_{\ell} c_\ell \bar{w}_\ell \tag{4.10}$$

for every $k \in \mathcal{K}$. Thus, under the latter assumption, the backward migration matrix M is not constant, and hence (4.6) must be supplemented by (4.10).

If migration precedes selection, we have the juvenile-migration model; see Nagylaki (1992, pp. 143–144) and references therein. These three migration models are qualitatively similar (their DC and CC approximations are identical), but quantitatively different. We focus on soft selection and relegate hard selection and juvenile migration to occasional remarks.

The special case without dominance is of particular biological interest and is often more tractable. This means that there exist constants $v_{i,k}$ such that

$$w_{ij,k} = v_{i,k} + v_{j,k} \tag{4.11}$$

for every $i, j \in \mathcal{J}$ and every $k \in \mathcal{K}$. Substituting (4.11) into (4.5) yields

$$w_{i,k} = v_{i,k} + \bar{v}_k, \qquad \bar{w}_k = 2\bar{v}_k, \tag{4.12a}$$

where

$$\bar{v}_k(p^{(k)}) = \sum_i v_{i,k} p_{i,k}, \tag{4.12b}$$

so (4.6) becomes

$$p'_{i,k} = \sum_\ell m_{k\ell} p_{i,\ell} (v_{i,\ell} + \bar{v}_\ell) / (2\bar{v}_\ell). \tag{4.13}$$

Note that $\bar{v}_k$ is linear, whereas $\bar{w}_k$ is quadratic.

4.2.2 General Results

After some introductory remarks, we examine five topics: the existence of an internal equilibrium, fixation of a specific allele, weak and strong migration, and uniform selection. The first problem is treated in NL1; the remainder are from NL7a.

4.2.2a General Comments

Even for two diallelic demes, convergence of (4.6) has not been proved; as we shall discuss in Sect. 4.3.2, for three diallelic demes, the gene frequencies need not converge. Only in very special cases can the equilibria of (4.6) be determined analytically.

In the diallelic case, most investigations have concerned the conditions for protecting an allele from loss, i.e., sufficient conditions for bounding its frequency away from 0 (Karlin, 1982; Nagylaki, 1992, Ch. 6; and references therein). These conditions ensure instability of the pertinent vertex equilibrium; they require that the spectral radius of an explicit nonnegative matrix exceed 1. If both alleles are protected, we have a protected polymorphism. This occurs if selection changes direction at least once (i.e., favors A_1 in some demes and A_2 elsewhere) and is sufficiently strong relative to migration. In addition to the above references, many results in this paper exemplify this fundamental conclusion.

The multiallelic case is far more difficult. Whereas $p_1 \in [0,1]^K$ fully describes a diallelic locus, for multiple alleles we require $p \in \Delta_J^K$. Since we can not locate the equilibria on the boundary $\partial \Delta_J^K$, we can not even derive useful protection conditions.

Some of our multiallelic results hold generically. We call a property generic if it holds for almost all fitnesses, migration matrices, and deme proportions, i.e., for an open, dense set of full measure.

4.2.2b Equilibrium

By Theorem 2.4 in NL1, if there is no dominance, the number of demes is a generic upper bound on the number of alleles present at equilibrium. By Remark 2.5 in NL1, this theorem is equally valid for hard selection.

The intuitive basis of this theorem is unclear. One might imagine that if an internal equilibrium exists, each allele must be the fittest in some deme, but Examples 4.6 and 4.7 in NL6b dispose of this idea. A slight modification of the proof of Theorem 2.4 in NL1 establishes the theorem for multiplicative fitnesses 2.4; i.e., instead of (4.11), we posit that there exist constants $v_{i,k}$ such that $w_{ij,k} = v_{i,k}v_{j,k}$ for every $i, j \in \mathcal{J}$ and every $k \in \mathcal{K}$. Now consider intermediate dominance, i.e., fitness schemes such that

$$\min(w_{ii,k}, w_{jj,k}) \leq w_{ij,k} \leq \max(w_{ii,k}, w_{jj,k}) \tag{4.14}$$

for every $i, j \in \mathcal{J}$ and every $k \in \mathcal{K}$. By Remark 4.4 in NL7a, the theorem does hold under (4.14) for sufficiently weak migration. However, Remark 4.15 in NL7a demonstrates that if migration is sufficiently strong, the theorem does not extend to (4.14).

Careful readers will realize that, since (4.6) need not converge, our theorem does not imply loss of an allele when $J > K$.

4.2.2c Fixation of an Allele

We invoke Remarks 3.2 and 3.3 in NL7a to paraphrase Proposition 3.1 in NL7a. Suppose that initially A_1 is present in some deme (i.e., $p_{1,\ell}(0) > 0$ for some $\ell \in \mathcal{K}$) and M is irreducible. Assume further that

$$w_{1i,k} + w_{1j,k} \geq 2w_{ij,k} \tag{4.15}$$

for every $i, j \in \mathcal{J}$ and every $k \in \mathcal{K}$, and that there exists some $k \in \mathcal{K}$ such that (4.15) is strict for either $j = 1$ or $j = i$ for every $i \in \mathcal{J}_1$. Then A_1 is ultimately fixed; i.e., $p_1(t) \to u$ as $t \to \infty$, where

$$u = (1, \ldots, 1)^{\mathrm{T}} \in \mathcal{R}^K. \tag{4.16}$$

Inequality (4.15) requires that the panmictic fixation condition in Proposition 2.16 in Nagylaki and Lou (2006a) hold in every deme. Therefore, (4.15) is quite strong. Can it be weakened?

4.2.2d Weak Migration

As already mentioned, the gene frequencies $p(t)$ do not always converge. However, under pure selection, they do (Lyubich et al., 1980; Losert and Akin, 1983). Under pure migration, the generic assumption that M is ergodic (i.e., irreducible and aperiodic) implies convergence. Hence, we expect convergence if migration is either sufficiently weak or sufficiently strong. Theorems 4.1 and 4.5 in NL7a tell us much more than that.

To study weak migration, we fix the fitnesses $w_{ij,k}$ for every $i, j \in \mathcal{J}$ and every $k \in \mathcal{K}$; set

$$m_{k\ell} = \delta_{k\ell} + \epsilon\mu_{k\ell} \tag{4.17}$$

for every $k, \ell \in \mathcal{K}$, where $\delta_{k\ell}$ represents the Kronecker delta ($\delta_{kk} = 1$ and $\delta_{k\ell} = 0$ if $k \neq \ell$); and let $\epsilon \to 0+$ with $\mu_{k\ell}$ fixed for every $k, \ell \in \mathcal{K}$. If $\epsilon = 0$, the recursion (4.6) reduces in each deme to the pure-selection mapping

$$p'_{i,k} = p_{i,k} w_{i,k}(p^{(k)}) / \bar{w}_k(p^{(k)}). \tag{4.18}$$

We posit that every equilibrium of (4.18) is hyperbolic. This assumption is generic. Suppose that $\epsilon > 0$ is sufficiently small.

Theorem 4.1 in NL7a comprises three parts, of which (a) and (b) are essentially in Karlin and McGregor (1972a,b); our main result is part (c):

(a) The set of equilibria $\Sigma_0 \subset \Delta_{\mathcal{J}}^K$ of (4.18) contains only isolated points, as does the set of equilibria $\Sigma_\epsilon \subset \Delta_{\mathcal{J}}^K$ of (4.6). As $\epsilon \to 0$, each equilibrium in Σ_ϵ converges to the corresponding equilibrium in Σ_0.

(b) In the neighborhood of each asymptotically stable equilibrium point in Σ_0, there exists exactly one equilibrium point in Σ_ϵ, and it is asymptotically stable. In the neighborhood of each unstable internal (i.e., in $\mathrm{int}\Delta_{\mathcal{J}}^K$) equilibrium point in Σ_0, there exists exactly one equilibrium point in Σ_ϵ, and it is unstable. In the neighborhood of each unstable boundary (i.e., in $\partial\Delta_{\mathcal{J}}^K$) equilibrium point in Σ_0, there exists at most one equilibrium point in Σ_ϵ, and if it exists, it is unstable.

(c) The solution $p(t)$ of (4.6) converges to one of the equilibrium points in Σ_ϵ.

The mean fitness $\bar{w}(p)$ measures the capacity of a population to survive and reproduce. Thus, $\bar{w}(p)$ is one of the most fundamental, important, and informative quantities in population genetics. For (4.18), the mean fitness $\bar{w}_k(p^{(k)})$ in each deme is monotone nondecreasing, and the single-generation change $\Delta\bar{w}_k = 0$ only at equilibrium (Mulholland and Smith, 1959; Scheuer and Mandel, 1959; Atkinson et al., 1960; Kingman, 1961). For (4.6), nonconvergence of $p(t)$ implies that a Lyapunov function can exist only under some restrictions.

We define the mean fitness for (4.6) as

$$\bar{w}(p) = \chi(\bar{w}_1(p^{(1)}), \dots, \bar{w}_K(p^{(K)})), \tag{4.19}$$

where χ is a strictly increasing function of $\bar{w}_k$ for every $k \in \mathcal{K}$. By Remark 4.3 in NL7a, if p is bounded away from the equilibria of (4.18), then $\Delta\bar{w}(p) > 0$ for sufficiently small $\epsilon > 0$. If p is close to an equilibrium, however, the mean fitness may decrease.

4.2.2e Strong Migration

When migration dominates selection, we expect rapid reduction to small quantities of the gene-frequency differences among demes and approximately panmictic evolution of suitably averaged gene frequencies. This panmictic evolution should be driven by suitably averaged fitnesses, which will provide a unique mean fitness for the entire population.

To express our intuition precisely, we fix M, posit that it is ergodic, take

$$w_{ij,k} = 1 + \epsilon r_{ij,k} \tag{4.20}$$

with $r_{ij,k}$ fixed for every $i, j \in \mathcal{J}$ and every $k \in \mathcal{K}$, and let $\epsilon \to 0+$.

The maximal eigenvalue 1 of the ergodic stochastic matrix M is simple and exceeds every other eigenvalue in modulus (Gantmacher, 1959, p. 80). The corresponding maximal left eigenvector $\alpha \in \text{int}\Delta_K$ satisfies

$$\alpha^{\mathrm{T}} M = \alpha^{\mathrm{T}} \tag{4.21}$$

and is the unique stationary distribution of the Markov chain with transition matrix M. Thus, in the absence of selection, if a gene is sampled from the population (with any distribution, e.g., from deme k), the probability that t generations ago its ancestral gene was in deme ℓ converges to α_ℓ as $t \to \infty$.

The vector α depends only on the relative migration rates. More precisely, if we replace $m_{k\ell}$ by $m_{k\ell}^* \equiv \theta m_{k\ell}$ for every $k \in \mathcal{K}$ and every $\ell \in \mathcal{K}_k$, where the constant $\theta > 0$ is independent of k and ℓ and satisfies

$$m_{kk}^* = 1 - \theta \sum_{\ell \in \mathcal{K}_k} m_{k\ell} \geq 0$$

for every $k \in \mathcal{K}$, then α is unaltered (Nagylaki, 1998).

A migration pattern is conservative if and only if it does not change the deme proportions c. In this case, and only in this case, we have $\alpha = c$ (Nagylaki, 1980). We shall see that all the strong-migration results in this subsection depend on M only through averages in which deme k has weight α_k. For conservative migration, the weighting is by c_k, and therefore all trace of population subdivision disappears.

We average $p_{i,k}$ over k with respect to α:

$$P_i = \alpha^{\mathrm{T}} p_i, \quad P = (P_1, \ldots, P_J)^{\mathrm{T}} \in \Delta_J. \tag{4.22}$$

Recalling (4.16), we define the gene-frequency deviations q from P:

$$q_{i,k} = p_{i,k} - P_i, \tag{4.23a}$$
$$q_i = p_i - P_i u \in \mathcal{R}^K, \tag{4.23b}$$
$$q^{(k)} = p^{(k)} - P \in \mathcal{R}^J, \tag{4.23c}$$
$$q = (q^{(1)}; \ldots; q^{(K)})^{\mathrm{T}} \in \mathcal{R}^{JK}. \tag{4.23d}$$

From (4.23b), (4.22), and (4.16) we get

$$\alpha^{\mathrm{T}} q_i = P_i - P_i = 0 \qquad (4.24)$$

for every $i \in \mathcal{J}$. To deduce the strong-migration limit, one must analyze (4.6) in terms of $(P, q)^{\mathrm{T}}$ rather than p.

Next, we introduce the average selection coefficients of $A_i A_j$, A_i, and the entire population:

$$\rho_{ij} = \sum_k \alpha_k r_{ij,k}, \qquad (4.25a)$$

$$\rho_i(P) = \sum_j \rho_{ij} P_j, \qquad (4.25b)$$

$$\bar{\rho}(P) = \sum_{i,j} \rho_{ij} P_i P_j. \qquad (4.25c)$$

To obtain the natural time scale, we scale generations by the selection intensity:

$$t = [\tau / \epsilon]. \qquad (4.26)$$

We shall see that the strong-migration (or weak-selection) limit of (4.6) is the simple panmictic system

$$\frac{\mathrm{d} P_i}{\mathrm{d}\tau} = P_i[\rho_i(P) - \bar{\rho}(P)], \qquad (4.27a)$$

$$q = 0. \qquad (4.27b)$$

We posit that every equilibrium of (4.27a) is hyperbolic, and $\epsilon > 0$ is sufficiently small. Then Theorem 4.5 in NL7a states the following:

(a) The set of equilibria $\Xi_0 \subset \Delta_{\mathcal{J}}^K$ of (4.27) contains only isolated points, as does the set of equilibria $\Xi_\epsilon \subset \Delta_{\mathcal{J}}^K$ of (4.6). As $\epsilon \to 0$, each equilibrium in Ξ_ϵ converges to the corresponding equilibrium in Ξ_0.
(b) In the neighborhood of each equilibrium point in Ξ_0, there exists exactly one equilibrium point in Ξ_ϵ. The stability of each equilibrium in Ξ_ϵ is the same as that of the corresponding equilibrium in Ξ_0, i.e., each pair is either asymptotically stable or unstable.
(c) The solution $p(t)$ of (4.6) converges to one of the equilibrium points in Ξ_ϵ.

Let λ_1 designate the nonunit eigenvalue of M with the largest modulus. By Remark 4.8 in NL7a, for every κ_1 such that $|\lambda_1| < \kappa_1 < 1$ and for every $t \geq \tilde{t} = (\ln \epsilon) / \ln \kappa_1$, we have

$$q(t) = O(\epsilon). \qquad (4.28)$$

Generically, we can take $\kappa_1 = |\lambda_1|$.

On account of (4.25) and (4.28), the approximate average fitnesses of $A_i A_j$, A_i, and the entire population are

$$\omega_{ij}(P) = 1 + \epsilon \rho_{ij}, \quad \omega_i(P) = 1 + \epsilon \rho_i(P), \quad \bar{\omega}(P) = 1 + \epsilon \bar{\rho}(P). \qquad (4.29)$$

Proposition 4.9 in NL7a informs us that if (4.28) holds and P is bounded away from the equilibria of (4.27a), then $\Delta \bar{\omega}(P) > 0$, i.e., the approximate mean fitness is increasing.

We define the exact mean fitness of the population as

$$\bar{w}(p) = \sum_k \alpha_k \bar{w}_k(p^{(k)}). \qquad (4.30)$$

Note that (4.30) is a special case of (4.19). We now paraphrase Theorem 4.12 in NL7a; this is a deeper result than Proposition 4.9. For $t \geq 2\tilde{t}$, we have

$$\Delta q = O(\epsilon^2). \qquad (4.31)$$

If (4.28) and (4.31) hold and P is bounded away from the equilibria of (4.27a), then $\Delta \bar{w}(p) > 0$.

By Remark 4.14, the approximate rate of increase of both $\bar{\omega}(P)$ and $\bar{w}(p)$ is equal to the genic variance of the panmictic population with averaged allelic frequencies P. This is the strong-migration analog of the asymptotic fundamental theorem of natural selection for multilocus selection (Nagylaki, 1993; Nagylaki et al., 1999)

The reader may find it helpful to peruse Example 4.16 in NL7a, where our strong-migration results are specialized to two diallelic demes.

4.2.2f Uniform Selection

We have just seen that strong migration leads to a panmictic limit. In our investigation of uniform selection, we seek sufficient conditions for similar conclusions, i.e., conditions under which there is no genetic indication of population structure. Our major results apply to the DC and CC models; it is desirable to extend them to (4.6), for which we have only the following local result and theorems for weak or strong migration.

With a slight abuse of notation, we posit that $w_{ij,k} = w_{ij}$ for every $i, j \in \mathcal{J}$ and every $k \in \mathcal{K}$. Suppose that every equilibrium of (4.18) is hyperbolic. We say that an equilibrium $\hat{p} \in \Delta_{\mathcal{J}}^K$ is uniform if $\hat{p}^{(k)} \in \Delta_J$ is independent of k. Theorem 5.1 in NL7a demonstrates that every uniform selection equilibrium of (4.18) is an equilibrium of (4.6) and has the same stability as under pure selection. By Remark 5.2, the ultimate rate of convergence to equilibrium is determined entirely by selection and is independent of migration.

The next two results hold for weak migration, as defined in Sect. 4.2.2d, and strong migration, as defined in Sect. 4.2.2e, respectively. In both cases, we posit that $p(0) \in \mathrm{int}\, \Delta_{\mathcal{J}}^K$. For weak migration, we suppose that every equilibrium of (4.18) is hyperbolic, and that (4.18) has a uniform, globally asymptotically stable equilibrium point $\hat{p}$. For strong migration, we suppose that M is ergodic, every equilibrium of (4.27a) is hyperbolic, and (4.27) has a uniform, globally asymptotically stable equilibrium point $\hat{p}$. Then Corollaries 5.11 and 5.12 in NL7a inform us that in both cases, $p(t) \to \hat{p}$ as $t \to \infty$.

4.2.3 Examples

Here, we illustrate the theory with a variety of diallelic and multiallelic examples that provide insight into the maintenance of polymorphism. Since the Levene (1953) model has been extensively studied, we examine it separately in Sect. 4.2.4.

4.2.3a Two Diallelic Demes

Without loss of generality, we put

$$w_{11,k} = 1 - r_k, \qquad w_{12,k} = 1, \qquad w_{22,k} = 1 - s_k, \qquad (4.32)$$

where $r_k \le 1$ and $s_k \le 1$ for $k = 1$ and 2. We write the backward migration matrix as

$$M = \begin{pmatrix} 1 - m_1 & m_1 \\ m_2 & 1 - m_2 \end{pmatrix}, \qquad (4.33)$$

where $0 < m_k < 1$ for $k = 1$ and 2. Set

$$\kappa = \frac{m_1}{s_1} + \frac{m_2}{s_2}. \qquad (4.34)$$

Then for every fixed (m_1, m_2), the allele A_1 is protected from loss in the hatched region in Fig. 4.1 (Maynard Smith, 1970; Bulmer, 1972; Nagylaki, 1992, Sect. 6.4). Notice that this condition is independent of r_k (because it is derived by linearization at $p_1 = 0$). Thus, A_1 is protected if either A_1A_2 is fitter than A_2A_2 in both demes, or A_1A_2 is fitter in one deme and less fit in the other ($s_1 s_2 < 0$) and the migration-selection ratio $\kappa < 1$.

Now assume that there is no dominance. Then $r_k = -s_k$ and $0 < |s_k| \le 1$ for $k = 1, 2$ in (4.32). From the preceding paragraph we obtain the results depicted in Fig. 4.2: (1) in Ω_0, allele A_1 is not protected, but A_2 is; (2) in Ω_1, allele A_1 is protected, but A_2 is not; and (3) in Ω_+, there is a protected polymorphism, which can not occur with panmixia. The hatched region is

$$\Omega_+ = \{(s_1, s_2) : s_1 s_2 < 0 \text{ and } |\kappa| < 1\}. \qquad (4.35)$$

Thus, for protection of both A_1 and A_2, selection must be in the opposite direction in the two demes and sufficiently strong relative to migration.

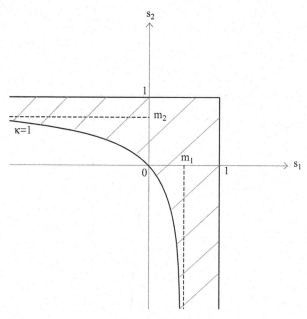

Fig. 4.1. The region of protection of A_1 (hatched). The parameters are defined in (4.32), (4.33), and (4.34)

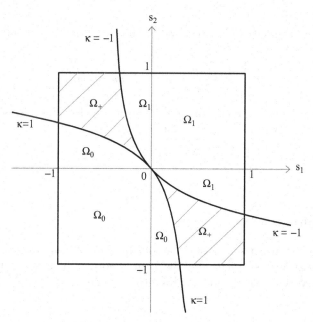

Fig. 4.2. The regions of protection of A_2 only (Ω_0), A_1 only (Ω_1), and both A_1 and A_2 (Ω_+) in the absence of dominance. See (4.32)–(4.35) for definitions

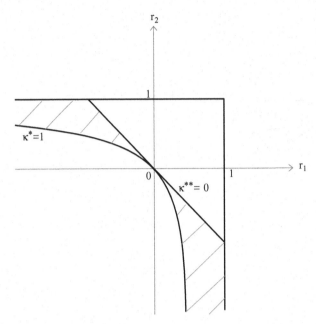

Fig. 4.3. The region of protected polymorphism (hatched) with A_1 recessive. See (4.36) and (4.37) for definitions

Finally, we posit that A_1 is recessive. Thus, we take $s_k = 0$ and $r_k \leq 1$ for $k = 1, 2$, and set

$$\kappa^* = \frac{m_1}{r_1} + \frac{m_2}{r_2}, \quad \kappa^{**} = m_2 r_1 + m_1 r_2. \tag{4.36}$$

Now the region of protected polymorphism, hatched in Fig. 4.3, is (Maynard Smith, 1970; Nagylaki, 1992, Sect. 6.4)

$$\Omega_- = \{(r_1, r_2) : r_1 r_2 < 0, \ \kappa^* < 1, \ \text{and} \ \kappa^{**} < 0\}. \tag{4.37}$$

The comment below (4.35) applies.

4.2.3b Multiple Alleles and Demes

We begin with an extremely simple, solvable model. Suppose that $J = K$, every heterozygote is lethal, and the homozygote $A_i A_i$ can survive only in deme i for every $i \in \mathcal{J}$. Therefore, the fitnesses are

$$w_{ij,k} = \beta_i \delta_{ij} \delta_{ik} \tag{4.38}$$

for some $\beta_i > 0$. From (4.38) and (4.5) we obtain

$$w_{i,k} = \beta_i \delta_{ik} p_{i,k}, \quad \bar{w}_k = \beta_k p_{k,k}^2. \tag{4.39}$$

Substituting (4.39) into (4.6) yields $p'_{i,k} = m_{ki}$, so

$$p_{i,k}(t) = m_{ki} \quad \text{for } t \geq 1. \tag{4.40}$$

The atypical, single-generation convergence is a consequence of lethality. In the solvable Example 2.8 in NL1, there is no dominance and convergence is geometric.

We now use the weak-migration theorem in Sect. 4.2.2d to extend Examples 2.1 and 2.2 in NL1 from local to global dynamics.

Generalizing Example 2.1 to multiple demes, we assume that $J \geq K$, and without migration, every equilibrium is hyperbolic and there exists a globally asymptotically stable equilibrium $\hat{p}$ such that, for each $k \in \mathcal{K}$,

$$\hat{p}_{i,k} > 0 \quad \text{if } i \in \mathcal{J}^{(k)}, \tag{4.41a}$$

$$\hat{p}_{i,k} = 0 \quad \text{otherwise}, \tag{4.41b}$$

where

$$\mathcal{J} = \bigcup_{k=1}^{K} \mathcal{J}^{(k)}, \quad \mathcal{J}^{(k)} \neq \emptyset, \quad \mathcal{J}^{(k)} \bigcap \mathcal{J}^{(\ell)} = \emptyset \tag{4.42}$$

for every $k, \ell \in \mathcal{K}$ such that $k \neq \ell$. Thus, when $p = \hat{p}$, each allele is present in exactly one deme. If migration is sufficiently weak, then $p(t) \to \tilde{p}$ as $t \to \infty$, where the equilibrium $\tilde{p}$ is close to $\hat{p}$. Furthermore, if M is irreducible, then $\tilde{p}_{i,k} > 0$ for every $i \in \mathcal{J}$ and every $k \in \mathcal{K}$, i.e., each allele is present in every deme.

In Example 2.2, we posited that $J \leq K$, there is no dominance, and every allele is the fittest in at least one deme:

$$\mathcal{K}^{(i)} = \{k \in \mathcal{K} : v_{i,k} > \max_{j: j \neq i} v_{j,k}\}, \quad \mathcal{K} = \bigcup_{i=1}^{J} \mathcal{K}^{(i)}, \tag{4.43}$$

$\mathcal{K}^{(i)} \neq \emptyset$, and $\mathcal{K}^{(i)} \cap K^{(j)} = \emptyset$ for every $i, j \in \mathcal{J}$ such that $i \neq j$. Without migration, the equilibrium $\hat{p}$, where $\hat{p}_{i,k} = 1$ if $i \in \mathcal{K}^{(i)}$ and $\hat{p}_{i,k} = 0$ if $i \notin \mathcal{K}^{(i)}$ for every $k \in \mathcal{K}$, is hyperbolic and globally asymptotically stable. This means that, for every $i \in \mathcal{J}$, ultimately A_i is fixed in $\mathcal{K}^{(i)}$ and eliminated from $\mathcal{K} - \mathcal{K}^{(i)}$. The last two sentences of the preceding paragraph apply unaltered.

For any model with strong migration, we suppose that M is ergodic and every equilibrium of (4.27a) is hyperbolic. Then $p(t) \to \tilde{p}$ as $t \to \infty$, where $\tilde{p}$ is an equilibrium of (4.6) that is close to the corresponding equilibrium $\hat{p}$ of (4.27), i.e., $\tilde{p}^{(k)}$ is close to an equilibrium $\hat{P}$ of (4.27a) for every $k \in \mathcal{K}$. If there are two demes with M given by (4.33), the definition (4.21) gives

$$\alpha = \frac{1}{m_1 + m_2} \begin{pmatrix} m_2 \\ m_1 \end{pmatrix}. \tag{4.44}$$

If there are two alleles and (4.32) holds, then (4.25a) yields (after scaling by ϵ)

$$\rho_{11} = -\sum_k \alpha_k r_k, \quad \rho_{12} = 0, \quad \rho_{22} = -\sum_k \alpha_k s_k. \tag{4.45}$$

4.2.4 The Levene Model

The Levene (1953) model is the simplest special case of (4.6). We formulate it, present general results, and offer examples in Sects. 4.2.4a, 4.2.4b, and 4.2.4c, respectively.

4.2.4a Formulation

Our formulation follows Nagylaki (1992, Sect. 6.3) and NL1.

The basic assumption of the Levene model is that individuals disperse independently of their deme of origin: $\tilde{m}_{k\ell} = \gamma_\ell$ for some positive constants γ_ℓ for every $k, \ell \in \mathcal{K}$. Then (4.9) immediately reveals that

$$m_{k\ell} = c_\ell^* \qquad (4.46)$$

for every $k, \ell \in \mathcal{K}$. This conclusion has two immediate implications. First, the Levene model has population subdivision but not geographical structure. According to (4.46), distance plays no role; in particular, migration probabilities do not decrease with separation. Second, (4.46) precludes a weak-migration limit.

Substituting (4.46) into (4.6) shows that, after one generation, the gene frequencies in zygotes are the same in every deme. With $p_{i,k} = p_i \in [0, 1]$ for every $i \in \mathcal{J}$ and every $k \in \mathcal{K}$, from (4.5) and (4.6) we obtain

$$w_{i,k}(p) = \sum_j w_{ij,k} p_j, \quad \bar{w}_k(p) = \sum_{i,j} w_{ij,k} p_i p_j, \qquad (4.47a)$$

$$p_i' = p_i \sum_k c_k^* w_{i,k} / \bar{w}_k. \qquad (4.47b)$$

In our treatment of (4.47), we simplify the notation from (4.4) to the definition

$$p = (p_1, \ldots, p_J)^{\mathrm{T}} \in \Delta_J. \qquad (4.48)$$

The classical interpretation of (4.47) is that intrademic selection is followed by random mating in the entire population.

For hard selection, (4.47) reduces to panmixia (Nagylaki, 1992, Sect. 6.3; NL1). Therefore, we focus on soft selection, for which (4.47b) becomes

$$p_i' = p_i \sum_k c_k w_{i,k} / \bar{w}_k. \qquad (4.49)$$

If there is no dominance, we insert (4.12a) into (4.49) to deduce the simpler model

$$\bar{v}_k(p) = \sum_i v_{i,k} p_i, \qquad (4.50a)$$

$$p_i' = \tfrac{1}{2} p_i \left(1 + \sum_k c_k \tfrac{v_{i,k}}{\bar{v}_k} \right). \qquad (4.50b)$$

4.2.4b General Results

Here, we reconsider, in the same order, the questions discussed in Sect. 4.2.2 for the general model (4.6), and we introduce some questions not treated there. Since (4.49) is much simpler than (4.6), we can provide a more detailed analysis.

An essential property of (4.49) is that the geometric-mean fitness

$$\tilde{w}(p) = \prod_k [\bar{w}_k(p)]^{c_k} \tag{4.51}$$

is nondecreasing (i.e., $\Delta \tilde{w} \geq 0$) and $\Delta \tilde{w} = 0$ only at equilibrium (Li, 1955; Cannings, 1971; Nagylaki, 1992, Sect. 6.3). This result implies generic convergence of $p(t)$ as $t \to \infty$.

It is often more convenient to study

$$F(p) = \ln \tilde{w}(p) = \sum_k c_k \ln \bar{w}_k(p) \tag{4.52}$$

instead of $\tilde{w}(p)$. Furthermore, by Lemma 3.2 in NL1, concavity of $\tilde{w}$ implies that of F, but not vice versa. Therefore, using F instead of $\tilde{w}$ can lead to stronger results. The main conclusions from this approach are Theorems 3.4 and 3.6 in NL1: If in every deme there is either no dominance or, without migration, a globally asymptotically stable, internal equilibrium, then $F(p)$ is concave. In this case, there exists exactly one stable equilibrium (point or manifold) and it is globally attracting. If there exists an internal equilibrium, it is globally asymptotically stable.

Next, we present some simple results for two alleles; consult Karlin (1977) for a much more detailed analysis. We write the fitnesses of $A_1 A_1$, $A_1 A_2$, and $A_2 A_2$ in deme k as x_k, 1, and y_k, respectively. Then A_1 is protected from loss if the harmonic mean of the fitnesses y_k is less than 1 (Levene, 1953; Nagylaki, 1992, Sect. 6.3):

$$y^* = \left(\sum_k c_k/y_k \right)^{-1} < 1. \tag{4.53}$$

Of course, A_2 is protected if

$$x^* = \left(\sum_k c_k/x_k \right)^{-1} < 1. \tag{4.54}$$

If A_1 is recessive, we have $y_k = 1$ for every $k \in \mathcal{K}$ and A_1 is protected if the arithmetic mean of x_k is greater than 1 (Prout, 1968; Nagylaki, 1992, Sect. 6.3):

$$\bar{x} = \sum_k c_k x_k > 1. \tag{4.55}$$

Consequently, there is a protected polymorphism if

$$x^* < 1 < \bar{x}. \tag{4.56}$$

This can not occur with panmixia.

To discuss these results, we define $\bar{y}$ as in (4.55) and set

$$x = (x_1, \ldots, x_K)^{\mathrm{T}} \in \mathcal{R}^K, \qquad y = (y_1, \ldots, y_K)^{\mathrm{T}} \in \mathcal{R}^K. \tag{4.57}$$

Note first that $y^* \leq \bar{y}$, with equality if and only if y_k is independent of k. Therefore, arithmetic-mean advantage of A_1A_2 over A_2A_2 ($\bar{y} < 1$) is sufficient but not necessary for protection of A_1.

By (4.53), allele A_1 is protected if A_2A_2 is sufficiently harmful in some deme, i.e., if there exists $k \in \mathcal{K}$ such that $y_k < c_k$.

If A_1 is recessive, from (4.56) we see that a sufficient condition for a protected polymorphism is that A_1 be sufficiently beneficial in some deme and sufficiently deleterious in some other deme, i.e., there exist $k, \ell \in \mathcal{K}$ such that $x_k > 1/c_k$ and $x_\ell < c_\ell$.

For given c and y, the harmonic mean y^* is minimized when c_k and $1/y_k$ have the same order (e.g., if we order the demes so that $c_1 \geq c_2 \geq \cdots \geq c_K$, then $y_1 \leq y_2 \leq \cdots \leq y_K$). Thus, protection of A_1 is most likely when it is favored in the larger demes. The same conclusion holds if A_1 is recessive because, for given c and x, the arithmetic mean $\bar{x}$ is maximized when c_k and x_k have the same order.

Using the results in Sect. 3.2 in NL1 on concavity of $F(p)$, Bürger (personal communication) proved that if for every $k \in \mathcal{K}$, either

$$x_k y_k \leq 1 + (1 - y_k)^2 \quad \text{and} \quad y_k \leq 1, \tag{4.58a}$$

or

$$x_k y_k \leq 1 + (1 - x_k)^2 \quad \text{and} \quad x_k \leq 1, \tag{4.58b}$$

then there exists exactly one stable equilibrium and it is globally attracting. Furthermore, if there exists an internal equilibrium, it is globally asymptotically stable. This theorem slightly weakens Karlin's (1977) sufficient condition $x_k y_k \leq 1$. We sketch (4.58) in Fig. 4.4.

We now turn to the multiallelic equilibrium problem. Its difficulty is suggested by Examples 3.10–3.12 in NL1, which exhibit the loss of genetic variability due to migration. The dynamics in these examples can not occur with two alleles. In Example 3.10, migration can cause global fixation of an allele that, without migration, is eliminated in every deme. In Example 3.12, migration can cause global fixation even if, without migration, there is a globally asymptotically stable, internal equilibrium in every deme.

By Proposition 3.13 in NL1, if there is no dominance, the number of demes is a generic upper bound on the number of alleles present at equilibrium.

Only in one case do we have a general solution for the equilibrium. Suppose that $J = K$ and there is no dominance. We introduce the $J \times J$ matrix V

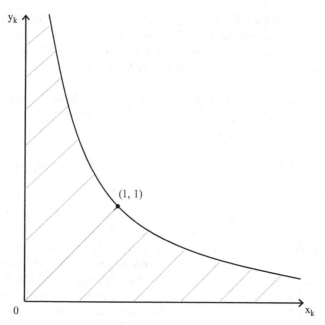

Fig. 4.4. The region of concavity (hatched) of F for the diallelic Levene model. See (4.52) and (4.58)

such that $(V)_{ki} = v_{i,k}$. Thus, row k and column i of V correspond to deme k and allele A_i, respectively. Under the generic assumption that $\det V \neq 0$, we have

$$V^{-1} = (\operatorname{adj} V)/\det V, \qquad (4.59)$$

where $\operatorname{adj} V$ represents the adjugate matrix of V. Let

$$d_k = \sum_j (\operatorname{adj} V)_{jk}, \qquad (4.60)$$

$$\hat{p}_i = \sum_k \tfrac{c_k}{d_k} (\operatorname{adj} V)_{ik}. \qquad (4.61)$$

By Theorem 4.2 in NL6b, an isolated, internal equilibrium point exists if and only if (4.61) satisfies $\hat{p}_i > 0$ for every $i \in \mathcal{J}$, in which case $\hat{p}$ is globally asymptotically stable.

We now discuss the loss of an allele. In discrete time, only in the Levene model do we have sufficient conditions for loss. Instead of examining (4.49), we investigate the more general mapping $\Delta_J \to \Delta_J$:

$$p_i' = p_i f_i(p) \quad \forall i \in \mathcal{J}, \qquad (4.62)$$

where

$$f_i(p) > 0 \qquad \forall p \in \operatorname{int}\Delta_J \text{ and } \forall i \in \mathcal{J}, \qquad (4.63a)$$

$$\sum_i p_i f_i(p) = 1 \quad \forall p \in \Delta_J, \qquad (4.63b)$$

and $f_i(p)$ is continuous for every $p \in \Delta_J$ for every $i \in J$. Although this recursion has many applications (NL6b, Sect. 3), it can not cover models with mutation, recombination, or arbitrary migration. The results described here greatly generalize parts of Sect. 2 in Nagylaki and Lou (2006a), which apply to pure selection.

Suppose that there exist vectors $a \in \Delta_J$ and $b \in \Delta_J$ such that

$$\sum_j a_j f_j(p) > \sum_j b_j f_j(p) \tag{4.64}$$

for every $p \in \text{int}\Delta_J$. For any vector $a \in \Delta_J$, we define the support of a as

$$S_a = \{i \in J : a_i > 0\}. \tag{4.65}$$

We say allele A_i is in S_a if $i \in S_a$. By Theorem 3.1 in NL6b, if (4.64) holds, then (4.62) has no internal equilibrium and there exist a and b such that $S_a \cap S_b = \emptyset$. We conjecture that the converse of this theorem does not hold.

This theorem implies that if $p(t)$ converges as $t \to \infty$, at least one allele must be eliminated. A more specific result follows from the additional assumptions that all the "favored" alleles (i.e., those in S_a) are initially present, and that (4.64) holds whenever all the "deleterious" alleles (i.e., those in S_b) are present. To state these hypotheses precisely, for any set $\mathcal{D} \subset J$, we define

$$\hat{\Delta}_J(\mathcal{D}) = \{p \in \Delta_J : p_j > 0 \ \forall j \in \mathcal{D}\} \tag{4.66}$$

and posit that there exist $a \in \Delta_J$ and $b \in \Delta_J$ such that (4.64) holds for every $p \in \hat{\Delta}_J(S_b)$. By Theorem 3.3 in NL6b, if $p(0) \in \hat{\Delta}_J(S_a)$ and $p(t)$ converges as $t \to \infty$, then $p_i(t) \to 0$ as $t \to \infty$ for some $i \in S_b$.

If $S_a = \{i\}$ for some $i \in J$ and we are satisfied with less detailed results, then we can delete the convergence assumption from the last theorem. We set $\tilde{\Delta}_J = \hat{\Delta}_J(\{i\} \cup S_b)$,

$$H_i(p) = \ln p_i - \sum_{j \in S_b} b_j \ln p_j, \tag{4.67}$$

$$g = \prod_{j \in S_b} p_j. \tag{4.68}$$

By Theorem 3.7 in NL6b, if $p(0) \in \tilde{\Delta}_J$, and

$$f_i(p) > \sum_{j \in S_b} b_j f_j(p) \tag{4.69}$$

for some $i \in J$, some $b \in \Delta_J$, and every $p \in \tilde{\Delta}_J$, then $H_i(p)$ is strictly increasing along orbits in $\tilde{\Delta}_J$, and $g(t) \to 0$ as $t \to \infty$. Note, however, that without convergence, this conclusion does *not* imply that $p_j(t) \to 0$ as $t \to \infty$ for some $j \in S_b$ (Akin and Hofbauer, 1982; Nagylaki, 1983; Akin and Losert, 1984).

Comparing (4.62) with (4.49), we see that with the identification

$$f_i(p) = \sum_k c_k w_{i,k}(p)/\bar{w}_k(p) \tag{4.70}$$

all the loss theorems apply to (4.49). In the generic case that all the equilibria of (4.49) are isolated, no convergence assumption is required. In Remark 3.11 in NL6b, we used (4.70) to replace (4.64) and (4.69) by slightly stronger but more explicit conditions in terms of allelic and genotypic fitnesses.

Of the many applications of the loss theorems in Sect. 3 in NL6b, we mention only one. By Remark 3.16 in NL6b, if in every deme, the homozygotes have the same fitness order and both underdominance and overdominance are absent, then the allele with the strictly greatest homozygote fitness in some deme is ultimately fixed.

Next, we examine fixation of an allele. Clearly, the theorem in Sect. 4.2.2c applies to (4.49). Corollary 3.15 in NL6b provides a set of sufficient conditions that is neither weaker nor stronger than (4.15): If

$$w_{ii,\ell} \geq w_{i+1,i+1;\ell} \quad \forall i \in \mathcal{J}_J, \ \forall \ell \in \mathcal{K}, \tag{4.71a}$$

$$w_{1i,\ell} \geq w_{ij,\ell} \quad \forall i,j \in \mathcal{J} \ni i \leq j, \ \forall \ell \in \mathcal{K}, \tag{4.71b}$$

and $w_{11,k} > w_{22,k}$ for some $k \in \mathcal{K}$, then $p_1(t) \to 1$ as $t \to \infty$. When there is no dominance, we have Proposition 4.1 in NL6b: If for some $i \in \mathcal{J}$,

$$\sum_k c_k v_{j,k}/v_{i,k} < 1 \tag{4.72}$$

for every $j \in \mathcal{J}_i$, then $p_i(t) \to 1$ as $t \to \infty$.

We continue with strong migration. Since $m_{k\ell} = c_\ell$, from (4.21) we obtain immediately that $\alpha = c$. Therefore, the strong-migration limit is (4.27) with $P = p$ and $q \equiv 0$. As observed in Remark 3.1 in NL1, this result can be derived directly from (4.49) and is just the weak-selection limit for a panmictic population.

We noted above (4.49) that hard selection reduces the Levene model to panmixia. Hence, the weak-selection limit is the same as for soft selection (NL1, Remark 3.1).

By censusing after migration, one can demonstrate that the weak-selection limit for juvenile migration (Nagylaki, 1992, pp. 143–144) is also panmixia. To prove this, one must follow the genotypic frequencies and establish that, after two generations, Hardy–Weinberg proportions hold with an error of $O(\epsilon^2)$.

Of course, for uniform selection, the Levene model simplifies at once to pure selection.

4.2.4c Examples

For the simple, solvable example (4.38), from (4.40) we get $p_i(t) = c_i$ for $t \geq 1$.

Henceforth, we posit that there is no dominance. For two diallelic demes with

$$v_{1,k} = \tfrac{1}{2}(1 + s_k), \quad v_{2,k} = \tfrac{1}{2}(1 - s_k), \tag{4.73}$$

where $0 < |s_k| \leq 1$ for $k = 1$ and 2, from (4.34) we obtain

$$\kappa = \frac{c_2}{s_1} + \frac{c_1}{s_2}. \tag{4.74}$$

Either (4.50b) or (4.61) yields the globally asymptotically stable, internal equilibrium

$$\hat{p}_1 = \tfrac{1}{2}(1 - \kappa), \quad \hat{p}_2 = \tfrac{1}{2}(1 + \kappa), \tag{4.75}$$

which exists if and only if $(s_1, s_2) \in \Omega_+$, defined in (4.35) and displayed in Fig. 4.2.

The rest of this section is devoted to multiple alleles with $J = K$. In Example 3.15 in NL1, we suppose that $v_{i,k} = \zeta_i \delta_{ik}$ for some $\zeta_i > 0$. Thus, allele A_i contributes to fitness only in deme i. Then we have

$$p_i(t) = c_i + [p_i(0) - c_i](\tfrac{1}{2})^t \to c_i \tag{4.76}$$

as $t \to \infty$.

To generalize greatly the preceding example, we assume that

$$v_{i,k} = \xi_k \eta_i + \nu_i \delta_{ik}, \tag{4.77}$$

where $\xi_k > 0$, $\eta_i \geq 0$, and $\nu_i > 0$ for every $i, k \in \mathcal{J}$. Thus, ξ_k modulates the intensity of panmictic selection in deme k with relative fitness contributions η_i, on which is superimposed a fitness increment ν_i for allele A_i in deme i.

Rewriting (4.77) as

$$v_{i,k} = \xi_k(\eta_i + \zeta_i \delta_{ik}), \tag{4.78}$$

where $\zeta_i = \nu_i / \xi_i$, we see from (4.50b) that the scale ξ_k cancels out and we have precisely Example 4.5 in NL6b. Consequently, there exists a globally asymptotically stable, internal equilibrium $\hat{p}$ if and only if

$$\frac{c_i \zeta_i(1 + \psi)}{1 + \psi - \eta_i} > \sum_j \frac{c_j \eta_j}{1 + \psi - \phi \eta_j} \tag{4.79a}$$

for every $i \in \mathcal{J}$, where

$$\phi = \sum_j \frac{1}{\zeta_j}, \quad \psi = \sum_j \frac{\eta_j}{\zeta_j}; \tag{4.79b}$$

in this case, we have

$$\hat{p}_i = \frac{c_i(1 + \psi)}{1 + \psi - \phi \eta_i} - \frac{1}{\zeta_i} \sum_j \frac{c_j \eta_j}{1 + \psi - \phi \eta_j}. \tag{4.80}$$

We discussed the solution (4.80) extensively in NL6b, where we derived it ab initio. It can also be proved by invoking Theorem 8.3.3 in Graybill (1969) to evaluate V^{-1} and then using (4.61).

The reader will find many other examples in Sects. 3.4 and 3.5 in NL1 and in Sect. 4.2 in NL6b.

4.3 Discrete Space and Continuous Time

In Sects. 4.3.1, 4.3.2, 4.3.3, and 4.3.4 we treat the formulation of the DC model, convergence, loss of an allele, and uniform selection, respectively.

4.3.1 Formulation

We now approximate (4.6) for weak evolutionary forces as in NL7a. We assume that

$$w_{ij,k} = 1 + \epsilon r_{ij,k}, \quad \tilde{m}_{k\ell} = \delta_{k\ell} + \epsilon \tilde{\mu}_{k\ell}, \tag{4.81}$$

where $r_{ij,k}$ and $\tilde{\mu}_{k\ell}$ are fixed for every $i, j \in \mathcal{J}$ and every $k \in \mathcal{K}$, and $\epsilon \to 0+$. From (4.5) and (4.81) we deduce

$$w_{i,k}(p^{(k)}) = 1 + \epsilon r_{i,k}(p^{(k)}), \quad \bar{w}_k(p^{(k)}) = 1 + \epsilon \bar{r}_k(p^{(k)}), \tag{4.82a}$$

in which

$$r_{i,k}(p^{(k)}) = \sum_j r_{ij,k} p_{j,k}, \quad \bar{r}_k(p^{(k)}) = \sum_{i,j} r_{ij,k} p_{i,k} p_{j,k}. \tag{4.82b}$$

To approximate M, note that (4.10) and (4.82a) imply that, for both soft and hard selection,

$$c_k^* = c_k + O(\epsilon) \tag{4.83}$$

as $\epsilon \to 0$. In fact, our results hold whenever (4.81) and (4.83) do. Substituting (4.81) and (4.83) into (4.9) leads to

$$m_{k\ell} = \delta_{k\ell} + \epsilon \mu_{k\ell}, \tag{4.84a}$$

where

$$\mu_{k\ell} = \frac{1}{c_k} \left(c_\ell \tilde{\mu}_{\ell k} - \delta_{k\ell} \sum_n c_n \tilde{\mu}_{nk} \right) + O(\epsilon) \tag{4.84b}$$

as $\epsilon \to 0$.

Since $\tilde{M}$ is stochastic, we have $\tilde{\mu}_{k\ell} \geq 0$ for every $\ell \in \mathcal{K}_k$ and

$$\sum_\ell \tilde{\mu}_{k\ell} = 0 \tag{4.85}$$

for every $k \in \mathcal{K}$. As required by the stochasticity of M, the leading term in (4.84b) satisfies the corresponding relations: for every $k \in \mathcal{K}$, we have $\mu_{k\ell} \geq 0$ for every $\ell \in \mathcal{K}_k$ and

$$\sum_\ell \mu_{k\ell} = 0. \tag{4.86}$$

Finally, we scale time and rename the gene frequencies:

$$t = [\tau/\epsilon], \quad \pi_{i,k}(\tau) = p_{i,k}(t). \tag{4.87}$$

Inserting (4.87), (4.84a), and (4.82a) into (4.6) and expanding yields

$$\pi_{i,k}(\tau + \epsilon) = \pi_{i,k}\{1 + \epsilon[r_{i,k}(\pi^{(k)}) - \bar{r}_k(\pi^{(k)})]\} + \epsilon \sum_\ell \mu_{k\ell}\pi_{i,\ell} + O(\epsilon^2) \quad (4.88)$$

as $\epsilon \to 0$, where

$$\pi^{(k)} = \pi^{(k)}(\tau) = (\pi_{1,k}(\tau), \ldots, \pi_{J,k}(\tau))^{\mathrm{T}} \in \Delta_J. \quad (4.89)$$

Rearranging (4.88) and letting $\epsilon \to 0$, we find

$$\frac{\mathrm{d}\pi_{i,k}}{\mathrm{d}\tau} = \sum_\ell \mu_{k\ell}\pi_{i,\ell} + \pi_{i,k}[r_{i,k}(\pi^{(k)}) - \bar{r}_k(\pi^{(k)})], \quad (4.90)$$

in which $\mu_{k\ell}$ is given by the leading term in (4.84b).

If we absorb ϵ into the migration rates and selection coefficients and return to $p(t)$, we obtain from (4.90) the slow-evolution approximation

$$\dot{p}_{i,k} = \sum_\ell \mu_{k\ell}p_{i,\ell} + p_{i,k}[r_{i,k}(p^{(k)}) - \bar{r}_k(p^{(k)})], \quad (4.91)$$

in which the superior dot signifies $\mathrm{d}/\mathrm{d}t$.

It is not difficult to show that (4.91) is also the slow-evolution approximation of the exact juvenile-migration model (Nagylaki, 1992, pp. 143–144).

4.3.2 Convergence

In Sects. 4.2.2d and 4.2.2e, we quoted theorems that prove convergence of (4.6) when migration is either weak or strong. In general, however, even the approximation (4.91) does not converge: Akin (personal communication) has established for three diallelic demes that Hopf bifurcation can produce unstable limit cycles. This result precludes global convergence, though not generic convergence.

With multiple alleles, there are no general results on the dynamics of (4.91). For two alleles, we set $p_k = p_{1,k}$ and write (4.91) in the form

$$\dot{p}_k = \sum_\ell \mu_{k\ell}p_\ell + \phi_k(p_k). \quad (4.92)$$

Since $\mu_{k\ell} \geq 0$ for every $\ell \in \mathcal{K}_k$, the system (4.92) is quasimonotone (Hirsch, 1982; Hadeler and Glas, 1983; Hofbauer and Sigmund, 1988, p. 158). By Theorem 3 in Hadeler and Glas (1983), the system (4.92) can not have an exponentially stable limit cycle. According to Theorem C in Hirsch (1982), if $K = 3$ and the ω-limit set $\mathcal{L}$ of (4.92) contains no equilibria, then $\mathcal{L}$ is a closed orbit. Theorem 2 in Hadeler and Glas (1983) informs us that if $K = 2$, then (4.92) converges as $t \to \infty$. Note that convergence of (4.6) has not been demonstrated even for two diallelic demes.

Eyland (1971) has provided a global analysis of (4.92) for two diallelic demes without dominance. We take $r_k = -s_k \neq 0$ in (4.32), $\mu_1 = \mu_{12} > 0$, and $\mu_2 = \mu_{21} > 0$. Then (4.92) becomes

$$\dot{p}_1 = \mu_1(p_2 - p_1) + s_1 p_1(1 - p_1), \tag{4.93a}$$
$$\dot{p}_2 = \mu_2(p_1 - p_2) + s_2 p_2(1 - p_2). \tag{4.93b}$$

Instead of (4.34), we have

$$\sigma_k = \frac{\mu_k}{s_k}, \qquad \kappa = \sigma_1 + \sigma_2. \tag{4.94}$$

Let $p = (p_1, p_2)^{\mathrm{T}}$. In the region Ω_0 in Fig. 4.2, allele A_1 is eliminated, i.e., $p(t) \to 0$ as $t \to \infty$, whereas in Ω_1, allele A_1 is ultimately fixed, i.e., $p(t) \to (1,1)^{\mathrm{T}}$ as $t \to \infty$. In Ω_+, $p(t)$ converges globally to an internal equilibrium point $\hat{p}$. If $s_2 < 0 < s_1$ and $|\kappa| < 1$, we have

$$\hat{p}_1 = \tfrac{1}{2}(1 + B) - \sigma_1, \qquad \hat{p}_2 = \tfrac{1}{2}(1 - B) - \sigma_2, \tag{4.95a}$$

where

$$B = (1 - 4\sigma_1\sigma_2)^{1/2}. \tag{4.95b}$$

4.3.3 Loss of an Allele

As remarked in Sect. 4.2.4b, we have sufficient conditions for global loss of an allele in discrete time only in the Levene model. For (4.91), however, we do have general conditions, which we proceed to discuss.

Suppose that there exist $i \in \mathcal{J}$ and constants γ_{ij} such that

$$\gamma_{ij} \geq 0, \qquad \gamma_{ii} = 0, \qquad \sum_j \gamma_{ij} = 1, \tag{4.96a}$$

$$\sum_j \gamma_{ij} r_{j,k}(p^{(k)}) > r_{i,k}(p^{(k)}) \tag{4.96b}$$

for every $k \in \mathcal{K}$ and every $p^{(k)} \in \Delta_J$ such that $p_{i,k} > 0$. We define

$$\Gamma_i = \{j \in \mathcal{J} : \gamma_{ij} > 0\} \tag{4.97a}$$

and posit that

$$p_i(0) > 0, \qquad p_j(0) > 0 \; \forall j \in \Gamma_i. \tag{4.97b}$$

By Theorem 3.5 in NL7a, if M is irreducible and (4.96) and (4.97) hold, then $p_i(t) \to 0$ as $t \to \infty$.

As explained in Remark 3.7 in NL7a, for some applications it is useful to replace (4.96b) by a slightly stronger assumption expressed directly in terms of genotypic selection coefficients: It suffices to posit that there exist constants γ_{ij} that satisfy (4.96a) and

$$\sum_j \gamma_{ij} r_{jh,k} > r_{ih,k} \tag{4.98}$$

for every $h \in \mathcal{J}$ and every $k \in \mathcal{K}$.

If there is no dominance, then (4.11) holds, so there exist constants $s_{i,k}$ such that

$$r_{ij,k} = s_{i,k} + s_{j,k} \tag{4.99}$$

for every $i, j \in \mathcal{J}$ and every $k \in \mathcal{K}$. Then both (4.96) and (4.98) simplify to the assumption that there exist $i \in \mathcal{J}$ and constants γ_{ij} that satisfy (4.96a) and

$$\sum_j \gamma_{ij} s_{j,k} > s_{i,k} \tag{4.100}$$

for every $k \in \mathcal{K}$. According to Corollary 3.9 in NL7a, if M is irreducible and (4.97), (4.99), and (4.100) hold, then $p_i(t) \to 0$ as $t \to \infty$.

To discuss the biological consequences of the preceding result, we set

$$s_{i,k}^* = \max_{j \in \mathcal{J}_i} s_{j,k}, \qquad s_{i,k}^{**} = \min_{j \in \mathcal{J}_i} s_{j,k}. \tag{4.101}$$

The averaging in (4.100) implies that $s_{i,k} < s_{i,k}^*$ for every $k \in \mathcal{K}$, i.e., in no deme is A_i the fittest allele. The elimination of A_i is obvious if $s_{i,k} < s_{i,k}^{**}$ for every $k \in \mathcal{K}$, but not if

$$s_{i,k}^{**} < s_{i,k} < s_{i,k}^* \tag{4.102}$$

for some $k \in \mathcal{K}$. Then in those demes (and possibly in every deme), the allele A_i has intermediate fitness, whereas some other allele has minimal fitness and another allele has maximal fitness. Thus, the intermediate allele is eliminated.

We can amplify this conclusion by specializing (4.100). If $\gamma_{i1} > 0$, $\gamma_{iJ} > 0$, and $\gamma_{ij} = 0$ for $j = 2, \ldots, J - 1$, then (4.100) becomes

$$\gamma_{i1} s_{1,k} + \gamma_{iJ} s_{J,k} > s_{i,k} \tag{4.103}$$

for every $k \in \mathcal{K}$. If (4.103) holds for $i = 2, \ldots, J - 1$, then $p_i(t) \to 0$ as $t \to \infty$ for $i = 2, \ldots, J - 1$.

Now, in addition to (4.103), suppose that

$$\min(s_{1,k}, s_{J,k}) < s_{i,k} < \max(s_{1,k}, s_{J,k}) \tag{4.104}$$

for $i = 2, \ldots, J - 1$ and every $k \in \mathcal{K}$. Thus, for $i = 2, \ldots, J - 1$, in each deme, the allele A_i is intermediate, whereas A_1 and A_J are extreme. Every intermediate allele is eliminated. As discussed on p. 394 of LN1, this conclusion can be interpreted as the elimination of generalists by specialists, and it can also yield the increasing phenotypic differentiation required for parapatric speciation. Finally, note that if $s_{1,k} - s_{J,k}$ changes sign as k varies, so that A_1

is the fittest allele in some deme(s) and A_J in the other deme(s), then both A_1 and A_J may be maintained in the population.

Of the many applications of the general loss theorem in Sect. 3 in NL7a, we mention only one. By Remark 3.18 in NL7a, if in every deme, the homozygotes have the same fitness order and there is strict heterozygote intermediacy, then the allele with the greatest homozygote fitness is ultimately fixed.

Lastly, we observe that Remark 3.20 in NL7a shows that (4.100) is sufficient but not necessary. A fortiori the same holds for (4.96). Can a weaker condition be found?

4.3.4 Uniform Selection

The results described in Sect. 4.2.2f for (4.6) with uniform selection cover local behavior and weak and strong migration. For (4.91), we have two theorems that guarantee global convergence to a uniform equilibrium point.

With a slight abuse of notation, we posit that $r_{ij,k} = r_{ij}$ for every $i, j \in \mathcal{J}$ and every $k \in \mathcal{K}$. As demonstrated by example in Sect. 5.2 in NL7a, this assumption does not suffice for global convergence to a uniform equilibrium. For uniform selection, the definitions (4.82b) reduce to

$$r_i(p^{(k)}) = \sum_j r_{ij} p_{j,k}, \quad \bar{r}(p^{(k)}) = \sum_{i,j} r_{ij} p_{i,k} p_{j,k}. \qquad (4.105)$$

From (4.91) we obtain the pure-selection system

$$\dot{p}_{i,k} = p_{i,k}[r_i(p^{(k)}) - \bar{r}(p^{(k)})]. \qquad (4.106)$$

We posit that (4.106) has a uniform, globally asymptotically stable, internal equilibrium point $\hat{p}$.

By Theorem 5.6 in NL7a, if the last two assumptions hold, $p(0) \in \mathrm{int}\,\Delta_J^K$, and all the migration rates are positive (i.e., $\mu_{k\ell} > 0$ for every $k \in \mathcal{K}$ and every $\ell \in \mathcal{K}_k$), then $p(t) \to \hat{p}$ as $t \to \infty$. We conjecture that this theorem remains valid even if $\hat{p} \in \partial \Delta_J^K$ and some migration rates are zero. By Theorem 5.9 in NL7a, the preceding theorem holds also if the migration rates are symmetric (i.e., $\mu_{k\ell} = \mu_{\ell k}$ for every $k, \ell \in \mathcal{K}$) rather than positive.

4.4 Continuous Space and Continuous Time

Discrete-space and diffusion (i.e., CC) models are complementary because their advantages and disadvantages are interchanged. As we have seen in Sects. 4.2 and 4.3, discrete-space models can be formulated and explored with fairly elementary mathematics, and they have some simple, relatively tractable special cases, such as two demes and the Levene model. However, general results are difficult to derive and not many have been proved. In contrast, the formulation and investigation of the diffusion model requires more advanced mathematics, and even its special cases are not easy. However, many interesting general results have been deduced. The diffusion model is particularly well suited to the study of stable, nonuniform gene-frequency equilibria, called clines. These are of great biological interest because they often enable us to establish the existence of selection, and if we have information about migration, to estimate the selection intensity (Haldane, 1948; Endler, 1977; Avise, 1994).

In this section, we review the diffusion theory of migration-selection models and present some examples.

In Sect. 4.4.1, we formulate the model for arbitrary (but genotype- and frequency-independent) migration. This level of generality is desirable because no fundamental principle imposes symmetries (e.g., homogeneity, isotropy) on migration. Since conditions are imposed on migration in much of the literature and in some of our results, these conditions are the subject of Sect. 4.4.2.

In Sect. 4.4.3, we treat the simplest case of two alleles, to which earlier work was confined. Even though the model is a scalar, semilinear parabolic equation, many difficult, open problems remain.

The remaining sections concern multiple alleles, described by a semilinear parabolic system. In Sects. 4.4.4 and 4.4.5, we examine the loss and protection of a particular allele. Section 4.4.6 is devoted to strong migration.

Many results for the important special case without dominance appear in Sects. 4.4.1–4.4.6; we investigate it further in Sect. 4.4.7. We devote Sect. 4.4.8 to uniform selection. Section 4.4.9 comprises our examples.

4.4.1 Formulation

We follow the formulation of the diffusion model in Sect. 1 in LN4; consult Sect. 1 in LN6 for a discussion of the various derivations of this model.

We start with (4.6) and (4.9) with weak soft selection and weak migration that satisfies the standard assumptions for a diffusion process. We denote position in the finite habitat Ω (a bounded, open domain in $\mathcal{R}^d$) by the vector $x = (x_1, \ldots, x_d)^{\mathrm{T}}$ and measure time, t, in generations. The population density at x is $\rho(x)$. Let $M_\alpha(x)$ and $V_{\alpha\beta}(x)$ designate the mean displacement in direction x_α and the covariance of the displacements in directions x_α and x_β per generation; these drift and diffusion coefficients form the vector $M(x)$ and the symmetric, positive definite matrix $V(x)$. As in Sects. 4.2 and 4.3, we consider a single locus with J alleles A_i, where $i \in \mathcal{J}$. Note that i and j refer to alleles, whereas α and β refer to spatial components.

Let $p_i(x, t)$ signify the frequency of A_i at position x at time t. For every $x \in \bar{\Omega}$ and $t \geq 0$, the vector $p(x, t)$ must satisfy

$$p(x,t) \in \Delta_J = \left\{ p \in \mathcal{R}^J : \ p_i \geq 0 \ \ \forall i \in \mathcal{J}, \ \sum_{j=1}^{J} p_j = 1 \right\}. \qquad (4.107)$$

To avoid the trivial case of essentially absent alleles, we suppose

$$\int_\Omega p_i(x,0)\,\mathrm{d}x > 0 \tag{4.108}$$

for every $i \in \mathcal{J}$.

The selection coefficient of the genotype $A_i A_j$ is the function $r_{ij}(x)$. Of course, $r_{ij}(x) = r_{ji}(x)$ for every $i,j \in \mathcal{J}$ and every $x \in \bar{\Omega}$. In LN4, we incorporated possible frequency dependence, i.e., we allowed r_{ij} to depend on both x and p. Here, for simplicity and consistency with Sects. 4.2 and 4.3, we include only spatial dependence. Let $r_i(x,p)$ and $\bar{r}(x,p)$ represent the selection coefficient of A_i and the mean selection coefficient of the population, respectively:

$$r_i(x,p) = \sum_{j=1}^{J} r_{ij}(x)p_j, \quad \bar{r}(x,p) = \sum_{i,j=1}^{J} r_{ij}(x)p_i p_j. \tag{4.109}$$

The contribution of selection is

$$S_i(x,p) = \lambda p_i(r_i - \bar{r}), \tag{4.110}$$

where λ (≥ 0) denotes the selection intensity (which we did not factor out in LN2). If time is scaled so that the migration rates are of order one, then λ becomes the ratio of the strength of selection to that of migration. Thus, weak, intermediate, and strong migration relative to selection correspond respectively to large, intermediate, and small values of λ compared with 1.

We define the divergence of an arbitrary symmetric matrix $W(x)$ as the vector with components ($\alpha = 1, 2, \ldots, d$)

$$(\nabla \cdot W)_\alpha = \sum_{\beta=1}^{d} W_{\alpha\beta, x_\beta}, \tag{4.111}$$

where the subscript x_β indicates partial differentiation. We introduce the vector

$$b(x) = \rho^{-1} \nabla \cdot (\rho V) - M \tag{4.112}$$

and the operators L and B defined by

$$Lu = \frac{1}{2} \sum_{\alpha,\beta=1}^{d} V_{\alpha\beta} u_{x_\alpha x_\beta} + b \cdot \nabla u, \tag{4.113a}$$

$$Bu = \nu \cdot V \nabla u, \tag{4.113b}$$

where ν denotes the unit outward normal vector on the boundary $\partial\Omega$.

The gene frequencies $p(x, t)$ satisfy the semilinear parabolic system (Nagylaki, 1975, 1989, 1996)

$$p_{i,t} = Lp_i + S_i(x, p) \quad \text{in } \Omega \times (0, \infty), \tag{4.114a}$$

$$Bp_i = 0 \quad \text{on } \partial\Omega \times (0, \infty), \tag{4.114b}$$

$$p(x, 0) \in \Delta_J \quad \text{in } \bar{\Omega} \tag{4.114c}$$

for every $i \in \mathcal{J}$. Here, L describes migration and (4.114b) specifies that no individuals cross the boundary. We are given $\rho(x)$, $M(x)$, $V(x)$, λ, $r_{ij}(x)$, and $p(x, 0)$; we seek the asymptotic behavior of $p(x, t)$ at $t \to \infty$.

The system (4.114) applies also to hard selection and juvenile migration (Nagylaki, 1989, 1996). If ploidy-weighted average gene-frequencies and selection, drift, and diffusion coefficients are used, then (4.114) holds for autosomal and X-linked loci in dioecious populations (Nagylaki, 1996). For plant populations with selfing and pollen and seed dispersal, appropriate definitions lead to the same migration operators L and B, but the selection term depends on the local selfing rate; if the latter is zero, then the selection term reduces to S_i (Nagylaki, 1997).

Throughout this paper, we assume that $\rho(x)$, $M_\alpha(x)$, $V_{\alpha\beta}(x)$, and $r_{ij}(x)$ are all Hölder continuous and $p_i(x, 0)$ is continuous in $\bar{\Omega}$. We assume also that $\partial\Omega \in \mathcal{C}^2$. By the maximum principle (Protter and Weinberger, 1984, Chap. 3) and the standard existence theory of evolution equations, the problem (4.114) has a unique classical solution $p(x, t)$ that exists for all time, $p_i \in \mathcal{C}(\bar{\Omega} \times [0, \infty)) \cap \mathcal{C}^{2,1}(\bar{\Omega} \times (0, \infty))$, and $0 < p_i(x, t) < 1$ for every $i \in \mathcal{J}$, every $x \in \bar{\Omega}$, and every $t > 0$. Therefore, without loss of generality, we posit that $0 < p_i(x, 0) < 1$ for every $i \in \mathcal{J}$ and every $x \in \bar{\Omega}$. This problem makes sense, i.e., (4.107) holds (LN2).

Some of our results depend on the positive eigenfunction ψ such that $L_1^\dagger \psi = 0$, where $L_1^\dagger$ designates the adjoint of the closure of L and B, discussed in more detail in Sect. 2.1 in LN2. In Sect. 4.4.2, we derive explicit formulas for ψ in special cases. We normalize $\psi(x)$ so that

$$\int_\Omega \psi(x) \, dx = 1. \tag{4.115}$$

If there is no dominance, then there exist Hölder-continuous selection coefficients $s_i(x)$ such that

$$r_{ij}(x) = s_i(x) + s_j(x) \tag{4.116}$$

for every $i, j \in \mathcal{J}$ and every $x \in \bar{\Omega}$. From (4.109) and (4.116) we obtain immediately

$$r_i(x, p) = s_i(x) + \bar{s}(x, p), \quad \bar{r}(x, p) = 2\bar{s}(x, p), \tag{4.117a}$$

where

$$\bar{s}(x, p) = \sum_{j=1}^{J} s_j(x) p_j. \tag{4.117b}$$

Inserting (4.117a) into (4.110) yields

$$S_i(x,p) = \lambda p_i(s_i - \bar{s}). \qquad (4.118)$$

Note that, although S_i in (4.110) is generally a cubic function of p, in (4.118) S_i is quadratic.

4.4.2 Particular Migration Patterns

Some of our results and examples require restrictions on migration. Here, we specialize L, B, and ψ for a hierarchy of restrictions.

The most general particularization of arbitrary migration is variational form, which applies if (4.113) can be written in the form

$$Lu = \frac{1}{c(x)}\, \nabla \cdot [A(x)\nabla u], \quad Bu = \nu \cdot A\nabla u, \qquad (4.119)$$

where $c(x)$ is a positive function, and $A(x)$ is a positive definite matrix for every $x \in \Omega$. By Theorem 4.1 in LN2, if (4.119) holds, then $\psi(x)$ is proportional to $c(x)$, and comparing (4.119) with (4.113) shows that $A = \frac{1}{2}cV$. Theorems 4.2 and 4.3 in LN2 provide necessary and sufficient conditions for (4.119).

In Theorem 4.4 in LN2, we established that each of the following five biologically transparent conditions suffices for validity of the variational form (4.119):

(a) *Unidimensionality*: Either the habitat Ω is actually one-dimensional or all the functions in the problem depend on only a single spatial variable.
(b) *Uniform migration*: M and V are independent of x in $\bar{\Omega}$.
(c) *Isotropic, uncorrelated migration*: $M(x) \equiv 0$ and $V(x) = V_0(x)I$, where $V_0(x)$ is a positive scalar function and I denotes the $d \times d$ identity matrix.
(d) *Conservative migration*: Migration does not change the population density $\rho(x)$ (Nagylaki, 1980).
(e) *Symmetric migration*: The underlying discrete migration pattern is described by a symmetric forward migration matrix, i.e., $\tilde{M}^{\mathrm{T}} = \tilde{M}$.

If (4.113) can be simplified to

$$Lu = \tfrac{1}{2}\nabla \cdot [V(x)\nabla u], \quad Bu = \nu \cdot V\nabla u, \qquad (4.120)$$

we say that L is in divergence form. Clearly, the choice $c(x) = 1$ and $A(x) = \frac{1}{2}V(x)$ reduces (4.119) to (4.120). Taking into account the normalization (4.115), we see at once that $\psi(x) = 1/|\Omega|$.

Divergence form applies if $\rho(x)$ is constant and (1) $M(x) \equiv 0$ and $V(x)$ is constant, (2) migration is conservative, or (3) migration is symmetric (LN4, Sect. 5.1).

Finally, if migration is homogeneous and isotropic, we can simplify (4.120) to

$$Lu = \nabla^2 u, \quad Bu = \nu \cdot \nabla u, \qquad (4.121)$$

the case studied in most of the literature.

4.4.3 Two Alleles

In this section, we specialize (4.114) to two alleles and present the most fundamental result of the theory. There is an extensive literature on the diallelic case; it can be traced from the many references in Nagylaki (1989, 1996, 1997), LN2, and LN6. There are many examples in Nagylaki (1975, 1976, 1978c), and we offer a few in Sect. 4.4.9a.

We simplify the notation by using p_1 instead of $(p_1, p_2)^T$ in $S_1(x, p)$ and setting

$$p(x, t) = p_1(x, t), \quad S(x, p) = S_1(x, p). \tag{4.122}$$

The condition (4.108) reduces to

$$0 < \frac{1}{|\Omega|} \int_\Omega p(x, 0) \, dx < 1. \tag{4.123}$$

From (4.109), (4.110), and (4.122) we obtain

$$S(x, p) = \lambda p(1 - p)\{r_{12}(x) - r_{22}(x) + [r_{11}(x) - 2r_{12}(x) + r_{22}(x)]p\}. \tag{4.124}$$

In general, $S(x, p)$ can not be factored into a function of x multiplied by a function of p. The factorization does hold, however, in the following biologically natural special cases.

(a) No dominance: Assumption (4.116) simplifies (4.124) to

$$S(x, p) = \lambda[s_1(x) - s_2(x)]p(1 - p). \tag{4.125}$$

(b) If A_1 is dominant, then $r_{11}(x) = r_{12}(x)$ for every $x \in \bar{\Omega}$ and (4.124) becomes

$$S(x, p) = \lambda[r_{11}(x) - r_{22}(x)]p(1 - p)^2. \tag{4.126}$$

(c) If A_1 is recessive, then $r_{12}(x) = r_{22}(x)$ for every $x \in \bar{\Omega}$ and (4.124) reduces to

$$S(x, p) = \lambda[r_{11}(x) - r_{22}(x)]p^2(1 - p). \tag{4.127}$$

(d) If all the selection coefficients have the same spatial dependence $g(x)$, we can choose (Nagylaki, 1975)

$$r_{11}(x) = g(x), \quad r_{12}(x) = \kappa g(x), \quad r_{22}(x) = -g(x), \tag{4.128}$$

in which κ signifies the degree of dominance. Substituting (4.128) into (4.124) gives

$$S(x, p) = \lambda g(x)p(1 - p)(1 + \kappa - 2\kappa p). \tag{4.129}$$

Observe that, with the obvious identifications of $g(x)$, the values 0, 1, and -1 of κ yield cases (a), (b), and (c), respectively.

Were κ a function of p, then (4.129) would generalize to the form studied by Fleming (1975) and Henry (1981):

$$S(x, p) = \lambda g(x) f(p). \tag{4.130}$$

The function $g(x)$ is Hölder continuous in $\bar{\Omega}$ and changes sign in Ω. The second assumption is crucial; it means that environmental diversity is reflected in at least one change in the direction of selection.

Following Henry (1981), we posit that $f \in C^2[0, 1]$,

$$f(0) = f(1) = 0, \quad f'(0) > 0 > f'(1), \tag{4.131a}$$
$$f''(p) \leq 0 \quad \text{in } [0, 1]. \tag{4.131b}$$

Here and below, primes indicate differentiation. If (4.129) applies, then (4.131) is equivalent to $|\kappa| \leq \frac{1}{3}$. Hence, (4.131) includes the case without dominance ($\kappa = 0$), but not complete dominance ($|\kappa| = 1$), underdominance ($|\kappa| > 1$), and overdominance ($|\kappa| > 1$).

From (4.114) and (4.130) we have the scalar problem

$$p_t = Lp + \lambda g(x) f(p) \quad \text{in } \Omega \times (0, \infty), \tag{4.132a}$$
$$Bp = 0 \quad \text{on } \partial\Omega \times (0, \infty), \tag{4.132b}$$
$$p(x, 0) \in (0, 1) \quad \text{in } \bar{\Omega}. \tag{4.132c}$$

To determine the asymptotic behavior of $p(x, t)$ as $t \to \infty$, we introduce the integral

$$C(g) = \int_\Omega g(x) \psi(x) \, dx. \tag{4.133}$$

Note that $q \equiv 1 - p$ satisfies (4.132) with $g(x)$ and $f(p)$ replaced by $-g(x)$ and $f^*(q) \equiv f(1 - q)$, respectively. Furthermore, f^* satisfies (4.131). Therefore, we may assume, without loss of generality, that $C(g) \leq 0$.

Now consider the linear eigenvalue problem

$$L\phi + \lambda f'(0) g(x) \phi = 0 \quad \text{in } \Omega, \tag{4.134a}$$
$$B\phi = 0 \quad \text{on } \partial\Omega. \tag{4.134b}$$

If $C(g) < 0$, then (4.134) has a unique positive eigenvalue λ_0 with a positive eigenfunction (Senn and Hess, 1982; LN6, p. 637).

We can now paraphrase Theorem 2.1 in LN2 as follows:

(a) Suppose $C(g) < 0$. If $0 < \lambda \leq \lambda_0$, then the trivial equilibrium $p = 0$ is asymptotically stable, and $p(x, t) \to 0$ uniformly in x as $t \to \infty$. If $\lambda > \lambda_0$, then (4.132) has a unique equilibrium $\hat{p}$ that satisfies $0 < \hat{p} < 1$. Furthermore, $\hat{p}$ is asymptotically stable and $p(x, t) \to \hat{p}(x)$ uniformly in x as $t \to \infty$.

(b) If $C(g) = 0$, then for every $\lambda > 0$, the problem (4.132) has a unique equilibrium $\tilde{p}$ that satisfies $0 < \tilde{p} < 1$. Furthermore, $\tilde{p}$ is asymptotically stable and $p(x, t) \to \tilde{p}(x)$ uniformly in x as $t \to \infty$.

This theorem extends the result of Henry (1981) from homogeneous, isotropic migration, in which (4.121) holds, to arbitrary migration. In part (a), when suitably averaged over the habitat, selection favors A_2, and a cline exists if and only if selection is sufficiently strong relative to migration. In part (b), the average selective forces are exactly balanced, and a cline always exists. These results agree qualitatively with those in Sects. 4.2 and 4.3. In contrast to the discrete models, however, here global convergence also has been proved.

Observe that the conditions for loss or protection of A_1 are monotone in λ. We shall see in Sect. 4.4.7c that this conclusion need not hold even for three alleles.

We now turn to some open problems, of which the most basic one is the relaxation of (4.131). We discussed this question in Sect. 5 in LN2, and conjectured that (4.131b) can be weakened to monotonicity of $f(p)/p$ for part (a) of the above theorem. For (4.129), proof of this conjecture would extend the interval of κ covered from $[-\frac{1}{3}, \frac{1}{3}]$ to $[-\frac{1}{3}, 1)$; if $f'(1) < 0$ in (4.131a) can be weakened to $f'(1) \leq 0$, then the interval becomes $[-\frac{1}{3}, 1]$.

For part (b) of the theorem, it might be possible to weaken (4.131b) to the uniqueness of the stationary point of $f(p)$ in $(0, 1)$. For (4.129), this would yield $|\kappa| < 1$; if (4.131a) is relaxed to $f'(0) \geq 0 \geq f'(1)$, we would obtain $|\kappa| \leq 1$.

If A_1 is recessive, as is common, then $\kappa = -1$ and hence

$$f(p) = 2p^2(1 - p). \tag{4.135}$$

For $C(g) < 0$, analogies and special cases (Nagylaki, 1975) suggest that there exists $\tilde{\lambda} > 0$ such that (1) if $\lambda < \tilde{\lambda}$, then $p = 0$ is globally asymptotically stable and (2) if $\lambda > \tilde{\lambda}$, then there exist exactly two nontrivial equilibria, $p^*(x)$ and $p^{**}(x)$, where 0 and p^{**} are asymptotically stable and 1 and p^* are unstable.

Much more work is required to assess the influence of ecological factors on clines. These include asymmetric migration, geographical barriers, and variation in the population density and migration rate. Consult Nagylaki (1975, 1976, 1978c, 1989, 1996, 1997), ten Eikelder (1979), Fife and Peletier (1981), Pauwelussen (1981), and Pauwelussen and Peletier (1981).

4.4.4 Loss of an Allele

After describing our general results, we specialize to the case without dominance.

4.4.4a General Results

We delineate sufficient conditions for loss of a specified allele. The main result is Theorem 1.1 in LN4, which generalizes Theorem 3.1 in LN2. As explained in Appendix B in LN6, the assumptions on which these two theorems are based must be slightly strengthened.

Our sufficient conditions are a slightly weaker, continuous-space version of (4.96). For any $i \in \mathcal{J}$ and sufficiently small $\delta > 0$, we define

$$\Delta_J^{(i)} = \{p \in \Delta_J : p_i = 0\}, \tag{4.136a}$$

$$\Delta_J^{(i,\delta)} = \{p \in \Delta_J : p_j \geq \delta \; \forall j \neq i\}. \tag{4.136b}$$

Suppose that there exist $i \in \mathcal{J}$ and constants γ_{ij} such that

$$\gamma_{ij} \geq 0, \quad \gamma_{ii} = 0, \quad \sum_j \gamma_{ij} = 1, \tag{4.137a}$$

$$\sum_j \gamma_{ij} r_j(x, p) \geq r_i(x, p) \tag{4.137b}$$

for every $x \in \bar{\Omega}$ and every $p \in \Delta_J$, and that there exists some $x \in \bar{\Omega}$ such that (4.137b) is strict for every $p \in \Delta_J^{(i)*}$, where

$$\Gamma_i = \{j \in \mathcal{J} : \gamma_{ij} > 0\}, \quad \Gamma_i^* = \{i\} \cup \Gamma_i, \tag{4.138a}$$

$$\Delta_J^{(i)*} = \{p \in \Delta_J : p_j > 0 \; \forall j \in \Gamma_i^*\}. \tag{4.138b}$$

Then Theorem 1.1 in LN4 and Appendix B in LN6 tell us that

(a) $p_i(x, t) \to 0$ uniformly in x as $t \to \infty$;
(b) if $p(x, 0) \in \Delta_J^{(i,\delta)}$, then

$$\max_{x \in \bar{\Omega}} p_i(x, t) \leq \frac{1}{\delta} \max_{x \in \bar{\Omega}} p_i(x, 0) \tag{4.139}$$

for every $t \geq 0$.

Part (a) of this theorem guarantees the elimination of A_i. Part (b) demonstrates that if $p(x, 0)$ is bounded away from every face of the simplex Δ_J other than $\Delta_J^{(i)}$, and if $p_i(x, 0)$ is small, then $p_i(x, t)$ remains small. Consequently, every equilibrium in int $\Delta_J^{(i)}$ that is (asymptotically) stable in $\Delta_J^{(i)}$ is (asymptotically) stable in Δ_J.

The biological discussion in Remarks 3.14–3.21 in NL7a applies here, as does that in Sect. 4.3.3 above. In particular, if for every $x \in \bar{\Omega}$, the homozygotes have the same fitness order and there is strict heterozygote intermediacy, then the allele with the greatest homozygote fitness is ultimately fixed (see Sect. 4.4.9b).

Sometimes it is convenient to replace (4.137) by a slightly stronger assumption expressed directly in terms of genotypic selection coefficients. Suppose that there exist $i \in \mathcal{J}$ and constants γ_{ij} that satisfy (4.137a) and

$$\sum_j \gamma_{ij} r_{jh}(x) \geq r_{ih}(x) \tag{4.140}$$

for every $h \in \mathcal{J}$ and every $x \in \bar{\Omega}$, and that there exists some $h \in \Gamma_i^*$ and some $x \in \bar{\Omega}$ such that (4.140) is strict. Multiplying (4.140) by p_h and summing over h leads to our previous assumption. However, since (4.137b) is strict only for $p \in \Delta_j^{(i)*}$, we can not take $p_h = 1$ in (4.137b) to derive the strict form of (4.140) for some $h \in \Gamma_i^*$.

For weak migration, Theorem 1.3 in LN4 gives more explicit sufficient conditions. We set

$$r^*(x, p) = \max_{j \in \mathcal{J}} r_j(x, p) \tag{4.141}$$

and posit that (1) there exists $i \in \mathcal{J}$ such that $r_i(x, p) < r^*(x, p)$ for every $x \in \bar{\Omega}$ and every $p \in \Delta_J$ and (2) for each $h \in \mathcal{J}$ and each $x \in \bar{\Omega}$, either $r_h(x, p) = r^*(x, p)$ for every $p \in \Delta_J$, or $r_h(x, p) < r^*(x, p)$ for every $p \in \Delta_J$. If λ is sufficiently large, then $p_i(x, t) \to 0$ uniformly in x as $t \to \infty$.

The intuitive assumption (1) means that A_i is always less fit than the fittest allele. Assumption (2) is technical; we expect that it is unnecessary.

4.4.4b No Dominance

We can simplify our results when there is no dominance. At the end of this section, we mention two other special cases.

From (4.117a) and (4.136) we deduce the sufficient condition for conclusions (a) and (b) above that there exist $i \in \mathcal{J}$ and constants γ_{ij} that satisfy (4.137a) and

$$\sum_j \gamma_{ij} s_j(x) \geq s_i(x) \tag{4.142}$$

for every $x \in \bar{\Omega}$ and that (4.142) is strict for some $x \in \bar{\Omega}$. This is Corollary 4.6 in LN4.

We now prove that for sufficiently small λ, condition (4.142) is not necessary. A fortiori, the same holds for (4.140). It is desirable to find a weaker loss condition. For two alleles and $i = 2$, assumption (4.142) simplifies to

$$s_1(x) \geq s_2(x) \tag{4.143}$$

for every $x \in \bar{\Omega}$, where (4.143) must be strict for some $x \in \bar{\Omega}$. We put

$$\sigma_i = \frac{1}{|\Omega|} \int_\Omega s_i(x) \, dx. \tag{4.144}$$

In Sect. 4.4.6, we shall see that if (4.120) holds, migration is sufficiently strong (i.e., λ is sufficiently small), and $\sigma_1 > \sigma_2$, then $p_1(x, t) \to 1$ uniformly in x as $t \to \infty$. Clearly, (4.143) can fail even if $\sigma_1 > \sigma_2$.

Assumption (4.116) simplifies also the weak-migration loss theorem. With the definition

$$s_i^*(x) = \max_{j \in \mathcal{J}_i} s_j(x), \tag{4.145}$$

we see from (4.117a) that assumption (2) below (4.141) always holds. There-fore, we have Corollary 4.7 in LN4: If there exists $i \in \mathcal{J}$ such that $s_i(x) < s_i^*(x)$ for every $x \in \bar{\Omega}$, and λ is sufficiently large, then $p_i(x, t) \to 0$ uniformly in x as $t \to \infty$. Example 4.8 in LN4 establishes that if the inequality is not strict, Corollary 4.7 fails.

With stronger assumptions, we can obtain more detailed results. If (4.142) holds for every $i \in \tilde{\mathcal{J}} \equiv \{2, 3, \ldots, J - 1\}$, then $p_i(x, t) \to 0$ uniformly in x as $t \to \infty$ for every $i \in \tilde{\mathcal{J}}$ and $p(x, t)$ converges globally to an equilibrium with either A_1 fixed (i.e., $p_1 = 1$), or A_J fixed, or both A_1 and A_J present (LN2, Sect. 3.3). Theorem 2.1 in LN2 (summarized above in Sect. 4.4.3) determines which equilibrium is the global attractor. The same results are proved under different assumptions in Theorem 3.2 in LN2; then Theorem 3.3 in LN2 spec-ifies the stability of all the vertices and diallelic equilibria (of course, there exist no other equilibria).

Instead of the additive selection coefficients in (4.116), one could posit multiplicative selection coefficients. This case is examined in Sect. 4.3 in LN4.

The genotype-independent spatial dependence in (4.130) can be gener-alized to multiple alleles; see Section 4.4 in LN4 for an exploration of the consequences of this assumption.

4.4.5 Protection of an Allele

We begin with our general results and then simplify them in the absence of dominance.

4.4.5a General Results

Before presenting our theorems on protection, we review some results of Senn and Hess (1982), as summarized in Sect. 1 in LN4.

For any continuous function $a(x) \not\equiv 0$, consider the linear eigenvalue problem

$$L\phi + \lambda a(x)\phi = 0 \quad \text{in } \Omega, \tag{4.146a}$$

$$B\phi = 0 \quad \text{on } \partial\Omega, \tag{4.146b}$$

and recall (4.133). If $a(x)$ is positive somewhere in $\bar{\Omega}$ and $C(a) < 0$, then (4.146) has a unique positive eigenvalue $\lambda_0(a)$ with a positive eigenfunction. If $C(a) \geq 0$, then (4.146) does not have a positive eigenvalue with a positive eigenfunction; in this case, we define $\lambda_0(a) = 0$. If $a(x) \leq 0$ for every $x \in \bar{\Omega}$, then (4.146) does not have a positive eigenvalue, and we define $\lambda_0(a) = \infty$.

We set

$$G_i(x) = \min_{j,h \in \mathcal{J}_1} [r_{ih}(x) - r_{jh}(x)] \tag{4.147}$$

and abbreviate $\beta_i = C(G_i)$. Observe that $G_i(x^{(i)}) > 0$ for some $x^{(i)} \in \bar{\Omega}$ if and only if

$$r_{ih}(x^{(i)}) > \max_{j \in \mathcal{J}_i} r_{jh}(x^{(i)}) \qquad (4.148)$$

for every $h \in \mathcal{J}_i$. By Proposition B in the Appendix to LN4, inequality (4.148) is equivalent to the statement that A_i is the fittest allele at $x^{(i)}$ when it is rare, i.e., in the limit $p_i \to 0$.

Protection of A_i means that there exists $\delta_i(\lambda) > 0$ such that, for all initial data that satisfy (4.108), there exists t_i (which may depend on λ and the initial data) such that $p_i(x,t) \geq \delta_i(\lambda)$ for every $x \in \bar{\Omega}$ and every $t \geq t_i$. Theorem 1.4, Remarks 1.6 and 1.7, and Proposition 4.1 in LN4 inform us that if $G_i(x^{(i)}) > 0$ for some $x^{(i)} \in \bar{\Omega}$, then A_i is protected (a) for every $\lambda > 0$ when $\beta_i \geq 0$ and (b) for every $\lambda > \lambda_0(G_i)$ when $\beta_i < 0$. Thus, if $G_i(x)$ is positive somewhere, then A_i is protected for sufficiently weak migration.

Our sufficient condition for protecting A_i requires that the entire face $\Delta_J^{(i)}$ of the simplex Δ_J be repelling. Consequently, if $J > 2$, it can be rather stringent. Can a weaker condition be derived?

We now turn to the existence of internal equilibria. According to Theorem 1.8 and Remark 4.4 in LN4, if for every $i \in \mathcal{J}$, there exists $x^{(i)} \in \bar{\Omega}$ such that $G_i(x^{(i)}) > 0$, then for every

$$\lambda > \max_{i \in \mathcal{J}} \lambda_0(G_i), \qquad (4.149)$$

the system (4.114) has at least one internal equilibrium. Thus, if every allele is protected by our criteria, then there exists at least one internal equilibrium.

We can now describe the limiting profiles of the internal equilibria. Let

$$H_i(x) = \min_{j \in \mathcal{J}_i}[r_{ii}(x) - r_{ji}(x)], \qquad (4.150)$$

$$\Omega_i = \{x \in \Omega : \ G_i(x) > 0, \ H_i(x) \geq 0\}. \qquad (4.151)$$

From Theorem 1.9 and Remark 4.4 in LN4, we see that if $\Omega_i \neq \emptyset$ for some $i \in \mathcal{J}$ and if $\hat{p}(x)$ is any internal equilibrium of (4.114), then $\hat{p}_i(x) \to 1$ uniformly in every compact subset of Ω_i as $\lambda \to \infty$. This is precisely the intuitively expected step-function weak-migration limit.

Consult Remarks 4.2, 4.3, and 4.5 in LN4 for more discussion.

4.4.5b No Dominance

In the absence of dominance, our results become considerably simpler. At the end of this section, we mention two other special cases.

From (4.147), (4.150), (4.151), (4.116), and (4.145) we find

$$G_i(x) = H_i(x) = s_i(x) - s_i^*(x), \qquad (4.152)$$

$$\Omega_i = \{x \in \Omega : \ s_i(x) > s_i^*(x)\}. \qquad (4.153)$$

By Corollary 4.9 in LN4, if $s_i(x^{(i)}) > s_i^*(x^{(i)})$ for some $i \in \mathcal{J}$ and some $x^{(i)} \in \bar{\Omega}$, then A_i is protected (a) for every $\lambda > 0$ when $\beta_i \geq 0$ and (b) for every $\lambda > \lambda_0(G_i)$ when $\beta_i < 0$. However, as explained on p. 313 in LN4, generically case (a) can apply to at most one allele.

Together, the weak-migration loss and protection corollaries (4.7 and 4.9, respectively, in LN4) demonstrate that, unless $s_i(x) \leq s_i^*(x)$ for every $x \in \bar{\Omega}$ and $s_i(x^*) = s_i^*(x^*)$ for some $x^* \in \bar{\Omega}$, the allele A_i is either globally eliminated or globally protected.

The results on internal equilibria (Corollary 4.10 in LN4) hold with the simplifications (4.152) and (4.153). As explained in Remark 4.11 in LN4, we suspect that for sufficiently large λ, there exists exactly one internal equilibrium and it is globally asymptotically stable.

For results on multiplicative selection coefficients and genotype-independent spatial dependence, see LN4, Sects. 4.3 and 4.4, respectively.

4.4.6 Strong Migration

Migration and selection act at rates of order 1 and λ, respectively. Therefore, if λ is sufficiently small, these time scales separate, and we expect migration to average the selection coefficients and gene frequencies so that the pure-selection system approximates (4.114). These ideas underlie the formal argument in Sect. 5.1 in LN4.

A local theorem in this direction is that of Carvalho and Hale (1991): if (4.120) holds, then as $\lambda \to 0$, every attractor of (4.114) converges to an attractor of the appropriate pure-selection system. We conjecture that, for arbitrary L, the diffusion analogue of the strong-migration Theorem 4.5 in NL7a, summarized in Sect. 4.2.2e, is valid. Such a result would describe all the equilibria and establish global convergence. Our global theorem, which we now present, is more restricted.

First, we replace $S_i(x, p)$ in (4.114a) by $T_i(x, p)$, which is Hölder continuous in x and Lipschitz in p but otherwise arbitrary. We posit that (4.120) holds and $p(x, t) \in \operatorname{int} \Delta_J$ for every $t > 0$ and every $x \in \bar{\Omega}$. Set

$$\mu = 1/\lambda, \qquad \tau = \lambda t, \qquad q(x, \tau) = p(x, t). \tag{4.154}$$

Then q satisfies the system

$$q_{i,\tau} = \mu L q_i + T_i(x, q) \quad \text{in } \Omega \times (0, \infty), \tag{4.155a}$$

$$B q_i = 0 \quad \text{on } \partial\Omega \times (0, \infty), \tag{4.155b}$$

$$q(x, 0) \in \operatorname{int} \Delta_J \quad \text{in } \bar{\Omega}. \tag{4.155c}$$

We define

$$\bar{T}_i(q) = \frac{1}{|\Omega|} \int_\Omega T_i(x, q) \, \mathrm{d}x \tag{4.156}$$

and consider the spatially averaged system

$$\frac{dq_i^*}{d\tau} = \bar{T}_i(q^*), \quad \tau > 0, \tag{4.157a}$$

$$q^*(0) \in \text{int } \Delta_J. \tag{4.157b}$$

We assume that the system (4.157) has a globally, linearly, asymptotically stable equilibrium $\hat{q}^*$. Then by Theorem 2.1 in LN6, for sufficiently large μ, the system (4.155) has a globally asymptotically stable equilibrium $\hat{q}(x)$, and $\hat{q}(x) \to \hat{q}^*$ uniformly in x as $\mu \to \infty$.

If $T_i(x, p)$ is independent of x for every $i \in J$, this theorem follows from a result of Conway et al. (1978).

We now apply the above theorem to the case without dominance. We recall (4.144), and with suitable labeling of the alleles, make the generic assumption that $\sigma_1 > \sigma_i$ for every $i \in J_1$. Still positing (4.120), we see that if λ is sufficiently small, then $p_1(x, t) \to 1$ uniformly in x as $t \to \infty$ (LN6, Theorem 1.1).

4.4.7 No Dominance

In the preceding sections, we deduced several results for the case without dominance by specializing general theorems. Here, we offer more detailed results from LN6 that we established directly. We treat the stability of vertices, that of edge equilibria, and the case of three alleles in Sects. 4.4.7a, 4.4.7b, and 4.4.7c, respectively.

4.4.7a Stability of Vertices

Theorem 1.5 in LN6 determines the stability of each vertex as follows. Recall (4.133), generalize (4.144) to

$$\sigma_i = C(s_i) = \int_\Omega s_i(x)\psi(x)\,dx, \tag{4.158}$$

and posit that

$$\sigma_1 > \max_{i \in J_1} \sigma_i. \tag{4.159}$$

Recalling the discussion of (4.146), we define

$$\lambda_1^* = \min_{i \in J_1} \lambda_0(s_i - s_1). \tag{4.160}$$

Suppose that $s_i(x) \not\equiv s_j(x)$ for every $i \in J$ and every $j \in J_i$, Then (a) every vertex other than vertex 1 is unstable; (b) vertex 1 is asymptotically stable if $\lambda < \lambda_1^*$; it is unstable if $\lambda > \lambda_1^*$.

Observe that if $s_i(x) \leq s_1(x)$ for every $i \in J_1$ and every $x \in \bar{\Omega}$, then $\lambda_1^* = \infty$, so vertex 1 is asymptotically stable for every $\lambda > 0$. If there exists $i \in J_1$ such that $s_i(x) > s_1(x)$ for some $x \in \bar{\Omega}$, then $0 < \lambda_1^* < \infty$, which implies that vertex 1 is asymptotically stable for sufficiently strong migration (or weak selection) and unstable for sufficiently weak migration (or strong selection). Note the consistency with the global theorem at the end of Sect 4.4.6.

4.4.7b Stability of Edge Equilibria

The next level of analysis is the examination of the existence and stability of the edge equilibria. We suppose throughout that $s_i(x) - s_j(x)$ changes sign in Ω for every i and j such that $1 \leq i < j \leq J$. This assumption is necessary for the existence of an equilibrium on the ij-edge.

Consider the problem $(1 \leq i < j \leq J)$

$$L\theta_{ij} + \lambda(s_i - s_j)\theta_{ij}(1 - \theta_{ij}) = 0 \quad \text{in } \Omega, \tag{4.161a}$$

$$B\theta_{ij} = 0 \qquad\qquad \text{on } \partial\Omega, \tag{4.161b}$$

$$0 < \theta_{ij}(x) < 1 \qquad \text{in } \Omega. \tag{4.161c}$$

According to Theorem 2.1 in LN2 (summarized in Sect. 4.4.3), there exists $\lambda_{ij} \geq 0$ such that (4.161) has a unique solution $\theta_{ij}(x)$ for every $\lambda > \lambda_{ij}$. Then the equilibrium $p^{(ij)}(x)$ of (4.114) on the ij-edge of Δ_J is given by

$$p_h^{(ij)}(x) = \begin{cases} \theta_{ij}(x) & \text{if } h = i, \\ 1 - \theta_{ij}(x) & \text{if } h = j, \\ 0 & \text{if } h \in \mathcal{J}_i \cap \mathcal{J}_j. \end{cases} \tag{4.162}$$

Let

$$s^{(ij)} = \max_{h \in \mathcal{J}_i \cap \mathcal{J}_j} s_h(x), \qquad \tilde{s}^{(ij)}(x) = \max[s_i(x), s_j(x)]. \tag{4.163}$$

The weak-migration Theorem 1.6 in LN6 informs us that for sufficiently large λ, the equilibrium $p^{(ij)}(x)$ is (a) asymptotically stable if $s^{(ij)}(x) < \tilde{s}^{(ij)}(x)$ for every $x \in \bar{\Omega}$; (b) it is unstable if there exists $x^{(ij)} \in \bar{\Omega}$ such that $s^{(ij)}(x^{(ij)}) > \tilde{s}^{(ij)}(x^{(ij)})$.

The stability condition here means that, for every $x \in \bar{\Omega}$, each allele absent at the edge equilibrium must be less fit than the fitter one of the two alleles present. In the DD model (4.13), the analog of this result follows directly from the theorem in Sect. 4.2.2d.

By Remark 3.3 in LN6, in part (a) above, the equilibrium $p^{(ij)}$ is globally asymptotically stable.

The next result determines the stability of each edge equilibrium immediately after its appearance as λ increases. We posit (4.159) and put

$$\tilde{\lambda}_{1h} = \min_{k \in \mathcal{J}_1 \cap \mathcal{J}_h} \lambda_{1k} \tag{4.164}$$

for every $h \in \mathcal{J}_1$. Theorem 1.7 in LN6 establishes the following:

(a) There exists $\epsilon_1 > 0$ such that $p^{(ij)}$ is unstable if $2 \leq i < j \leq J$ and $\lambda_{ij} < \lambda < \lambda_{ij} + \epsilon_1$.

(b) Suppose that $\lambda_{1h} < \tilde{\lambda}_{1h}$ for some $h \in \mathcal{J}_1$. Then there exists $\epsilon_2 > 0$ such that $p^{(1h)}$ is asymptotically stable if $\lambda_{1h} < \lambda < \lambda_{1h} + \epsilon_2$, and $p^{(1k)}$ is unstable if $k \neq h$ and $\lambda_{1k} < \lambda < \lambda_{1k} + \epsilon_2$.

Notice that generically there exists a unique $h \in \mathcal{J}_1$ that satisfies the assumption in part (b). To see this, suppose, without loss of generality, that $\lambda_{1k} \geq \lambda_{1,k+1}$ for every $k \in \mathcal{J}_1 \cap \mathcal{J}_J$. Then (4.164) yields $\tilde{\lambda}_{1J} = \lambda_{1,J-1} \geq \lambda_{1J}$ and $\tilde{\lambda}_{1h} = \lambda_{1J} \leq \lambda_{1h}$ for every $h \in \mathcal{J}_1 \cap \mathcal{J}_J$. If $\lambda_{1,J-1} > \lambda_{1J}$, then $h = J$ in part (b).

This theorem informs us that, as λ increases, the only edge equilibrium that is stable immediately after its emergence is the one that materializes first among the edge equilibria that involve the allele with the highest average fitness.

Can the stability of the edge equilibria be determined for arbitrary λ?

4.4.7c Three Alleles

In LN6 we proved that even in the triallelic model with homogeneous, isotropic migration, there exist complex, unexpected phenomena that can not occur in the diallelic case. Theorem 1.9 shows that as λ increases, arbitrarily many changes of stability of the edge equilibria and corresponding appearance of an internal equilibrium can occur. Furthermore, Theorem 1.10 and Remark 1.11 demonstrate that the condition for protection of a specified allele is not monotone in λ. Theorem 1.12 and Remark 1.13 establish the corresponding nonmonotonicity in λ of the condition for loss of a particular allele.

We deduced the above complicated properties under the assumption that two of the selection coefficients are arbitrarily close to each other for every $x \in \bar{\Omega}$. Perhaps these phenomena can not occur if the selection coefficients are bounded away from each other on a set of positive measure. Nonetheless, this complex behavior suggests the difficulty of a complete analysis of even the simplest multiallelic model. Consult the last two paragraphs in Sect. 6 in LN6 for open problems and more discussion.

4.4.8 Uniform Selection

In this section, we present the diffusion analogues of the local DD result in Sect. 4.2.2f and the global DC ones in Sect. 4.3.4. As in Sect. 4.3.4, with a slight abuse of notation, we posit that there exist constants r_{ij} such that $r_{ij}(x) = r_{ij}$ for every $i \in \mathcal{J}$, every $j \in \mathcal{J}$, and every $x \in \bar{\Omega}$. Then (4.109) reduces to

$$r_i(p) = \sum_j r_{ij} p_j, \quad \bar{r}(p) = \sum_{i,j} r_{ij} p_i p_j, \tag{4.165}$$

so the contribution of selection becomes

$$S_i(p) = \lambda p_i(r_i - \bar{r}). \tag{4.166}$$

The gene frequencies $p(x,t)$ satisfy (4.114) with $S_i(x,p)$ replaced by $S_i(p)$. In the absence of migration, we have

$$\frac{\mathrm{d}p_i}{\mathrm{d}t} = S_i(p). \tag{4.167}$$

At an equilibrium $\hat{p} \in \Delta_J$ of (4.167), we have $S(\hat{p}) = 0$. From (4.113) we see immediately that $\hat{p}$ is a uniform equilibrium of (4.114). By Theorem 2.1 in NL7b, if $\hat{p}$ is hyperbolic, then $\hat{p}$ has the same stability as under pure selection. Furthermore, according to Remark 2.3 in NL7b, if $\hat{p}$ is asymptotically stable, then the ultimate rate of convergence to $\hat{p}$ is determined entirely by selection and is independent of migration.

Our principal result on uniform selection is Theorem 3.3 in NL7b. If $\hat{p} \in$ int Δ_J is a globally asymptotically stable equilibrium point of (4.167), then $p(x,t) \to \hat{p}$ as $t \to \infty$. We conjecture that the same result holds for boundary equilibria.

Hadeler (1981; see also Redlinger, 1983) proved the special case of the last theorem for homogeneous, isotropic migration, i.e., under the assumption (4.121). The Lyapunov function we used to prove Theorem 3.3 in NL7b is different from his, which can be generalized from the Laplacian (4.121) to the variational form (4.119).

4.4.9 Examples

Below, we offer a few diallelic and multiallelic examples. Step-function selection coefficients should be interpreted as limits of Hölder-continuous functions.

4.4.9a Two Alleles

The diallelic form of the first loss theorem in Sect. 4.4.4 is very simple. Choosing $i = 1$ in (4.137), using p_1 instead of $(p_1, p_2)^{\mathrm{T}}$ in $r_i(x,p)$, and abbreviating $p = p_1$, we conclude that if

$$r_2(x,p) \geq r_1(x,p) \tag{4.168}$$

for every $x \in \bar{\Omega}$ and every $p \in [0,1]$, and if there exists some $x \in \bar{\Omega}$ such that (4.168) is strict for every $p \in (0,1)$, then $p(x,t) \to 0$ uniformly in x as $t \to \infty$. In terms of genotypic selection coefficients, this sufficient condition is equivalent to requiring that

$$r_{22}(x) \geq r_{12}(x) \geq r_{11}(x) \tag{4.169}$$

for every $x \in \bar{\Omega}$ and that at least one of the inequalities in (4.169) be strict for some $x \in \bar{\Omega}$. This is precisely (4.140).

For protection, note from (4.147) that

$$G_1(x) = r_{12}(x) - r_{22}(x), \qquad G_2(x) = r_{12}(x) - r_{11}(x). \tag{4.170}$$

The sufficient condition in the first theorem in Sect. 4.4.5 is equivalent to instability of the trivial equilibrium $p = 0$.

For other examples in the same spirit, see Remarks 4.2, 4.3, and 4.5 in LN4.

In Nagylaki (1975, 1976, 1978c), there is extensive discussion of qualitative properties of clines, space and time scales, and the influence of various ecological inhomogeneities. Of the many examples of step environments, we present only one. Suppose (4.116) and (4.121) hold; the habitat $\Omega = (-1, \gamma)$, where $\gamma > 0$;

$$g(x) = \begin{cases} 1 & \text{in } [-1, 0), \\ -\alpha^2 & \text{in } (0, \gamma]; \end{cases} \tag{4.171}$$

and

$$\int_{-1}^{\gamma} g(x)\, dx = 1 - \alpha^2 \gamma < 0. \tag{4.172}$$

From Nagylaki (1975) and the theorem in Sect. 4.4.3, we conclude that a globally asymptotically stable cline exists if $\lambda > \xi^2$, where ξ is the unique root of $(\alpha > 0)$

$$\xi = \tan^{-1}[\alpha \tanh(\alpha\gamma\xi)]. \tag{4.173}$$

Observe that $0 < \xi < \tan^{-1}\alpha < \frac{\pi}{2}$, and $\xi \to \tan^{-1}\alpha$ as $\gamma \to \infty$.

4.4.9b Multiple Alleles

The reader may find Examples 4.8 in LN4 and 1.4 in LN6 instructive.

We now establish a result of biological interest on fixation. Suppose that everywhere homozygotes have the same order and heterozygotes are intermediate. Then we can label the alleles so that

$$r_{ii}(x) \geq r_{ij}(x) \geq r_{jj}(x) \tag{4.174a}$$

for every $x \in \mathcal{J}$, every $j \in \mathcal{J}$ such that $i \leq j$, and every $x \in \bar{\Omega}$. Assume also that there exists $\tilde{x} \in \bar{\Omega}$ such that

$$r_{11}(\tilde{x}) > r_{22}(\tilde{x}). \tag{4.174b}$$

Then $p_1(x, t) \to 1$ uniformly in x as $t \to \infty$.

To prove this assertion, we employ the genotypic loss condition (4.140). From (4.174a) we obtain

$$r_{1i}(x) \geq r_{ii}(x) \geq r_{iJ}(x) \tag{4.175}$$

for every $i \in \mathcal{J}$ and every $x \in \bar{\Omega}$. Furthermore, (4.174) implies

$$r_{11}(\tilde{x}) > r_{22}(\tilde{x}) \geq r_{JJ}(\tilde{x}), \tag{4.176}$$

whence (4.174a) yields either $r_{11}(\tilde{x}) > r_{1J}(\tilde{x})$ or $r_{1J}(\tilde{x}) > r_{JJ}(\tilde{x})$. Therefore, $r_{1i}(\tilde{x}) > r_{iJ}(\tilde{x})$ for either $i = 1$ or $i = J$. We conclude that $p_J(x, t) \to 0$ uniformly in x as $t \to \infty$. Iterating this argument, we demonstrate successively that $p_i(x,t) \to 0$ uniformly in x as $t \to \infty$ for $i = J - 1, \ldots, 2$, which proves our assertion.

Finally, we present an application of the strong-migration theorem in Sect. 4.4.6. The corresponding DD example is in Remark 4.15 in NL7a. We assume that (4.120) holds and

$$\rho_{ii} = \tfrac{1}{|\Omega|} \int_\Omega r_{ii}(x)\, dx < 0, \tag{4.177a}$$

$$\rho_{ij} = \tfrac{1}{|\Omega|} \int_\Omega r_{ij}(x)\, dx = 0 \tag{4.177b}$$

for every $i \in \mathcal{J}$ and every $j \in \mathcal{J}_i$. Then the spatially averaged pure-selection system (4.157) has a globally, linearly, asymptotically stable equilibrium $\tilde{p}$ given by (Nagylaki, 1992, p. 98, Problem 4.14)

$$\tilde{p}_i = \frac{1}{\rho_{ii}} \bigg/ \sum_{j=1}^{J} \frac{1}{\rho_{jj}} > 0 \tag{4.178}$$

for every $i \in \mathcal{J}$. We infer that for sufficiently small λ, the system (4.114) has a globally asymptotically stable equilibrium $\hat{p}(x)$, and $\hat{p}(x) \to \tilde{p}$ uniformly in x as $\lambda \to 0$.

A few comments will illuminate this result.

First, it fails when there is no dominance. Indeed, from (4.177), (4.116), and (4.144) we get $\rho_{ii} = 2\sigma_i < 0$ and $\rho_{ij} = \sigma_i + \sigma_j = 0$ for every $i \in \mathcal{J}$ and every $j \in \mathcal{J}_i$, which is a contradiction.

Second, (4.174a) also precludes our result because (4.174a) and (4.177) yield $\rho_{ij} \leq \rho_{ii} < 0$ for every $i \in \mathcal{J}$ and every $j \in \mathcal{J}$ such that $i < j$. The contradiction rules out (4.174a).

Third, it is important to observe that, even though every allele is maintained in globally asymptotically stable equilibrium, assumption (4.177) requires neither underdominance nor overdominance. To see this, suppose that $\Omega = (-1,1)$ and

$$r_{ii}(x) < r_{ij}(x) < r_{jj}(x), \tag{4.179a}$$

$$r_{ij}(-x) = -r_{ij}(x), \tag{4.179b}$$

$$-r_{ij}(x) < r_{ii}(-x) < -r_{ii}(x) \tag{4.179c}$$

for every $i \in \mathcal{J}$, every $j \in \mathcal{J}$ such that $j > i$, and every $x \in (0,1]$. Note that $r_{ij}(x)$ is discontinuous at $x = 0$; we define $r_{ij}(0) = r_{ij}(0+)$.

From (4.179b,c) we deduce immediately that (4.177) holds. For $x \in (0,1]$, inequalities (4.179a) imply heterozygote intermediacy. Furthermore, (4.179b,c) give $r_{ij}(-x) < r_{ii}(-x)$. Successively using (4.179b), (4.179a), and (4.179c) leads to

$$r_{ij}(-x) = -r_{ij}(x) > -r_{jj}(x) > r_{jj}(-x) \tag{4.180}$$

for every $i \in \mathcal{J}$, every $j \in \mathcal{J}$ such that $j > i$, and every $x \in (0,1]$. Therefore, heterozygote intermediacy holds for every $x \in \Omega$.

We close with a concrete example in which the selection coefficients are step functions. We assume that

$$r_{ii}(x) = i, \quad r_{ij}(x) = i + \tfrac{1}{2}, \quad r_{ii}(-x) = -i - \tfrac{1}{4} \qquad (4.181)$$

for every $i \in \mathcal{J}$, every $j \in \mathcal{J}$ such that $j > i$, and every $x \in (0,1]$. Then (4.179) holds and $\rho_{ii} = -\tfrac{1}{8}$, whence (4.178) reduces to $\tilde{p}_i = 1/J$ for every $i \in \mathcal{J}$.

4.4.9c Acknowledgement

We are pleased to thank Ms. Liping Gao for drawing the figures and Prof. Reinhard Bürger for comments on the paper.

Appendix

In Table 4.1, we list and briefly define the symbols used in this paper. For both the Roman and Greek alphabets, uppercase letters precede lowercase ones. For each uppercase or lowercase letter, listing is in order of appearance of the precise definition in the text. The references are to the equation closest to the precise definition of each symbol. Thus, (7), (7)+, and (7)− would mean (7), the text below (7), and the text above (7), respectively. Note that some symbols signify different quantities in different sections.

Table 4.1. Glossary of symbols

Symbol	Reference	Definition
A_i	(4.1a)−	Alleles
A	(4.119)	$d \times d$ matrix in diffusion
a	(4.64)−	Constant vector in Δ_J
a	(4.146a)−	Scalar function of $x \in \bar{\Omega}$
a_i	(4.64)	Components of $a \in \Delta_J$
B	(4.95b)	Constant
B	(4.113b)	Boundary operator
b	(4.64)−	Constant vector in Δ_J
b	(4.112)	Vector in $\mathcal{R}^d$
b_i	(4.64)	Components of $b \in \Delta_J$
$\mathcal{C}^2$	(4.115)−	Indicates that the functions defining $\partial\Omega$ have continuous second partial derivatives
$\mathcal{C}$	(4.115)−	Space of continuous functions
$\mathcal{C}^{2,1}$	(4.115)−	Space of functions whose second spatial partial derivatives and first temporal partial derivative are continuous
C	(4.133)	Integral
c_k	(4.8)−	Pre-selection deme proportions
c_k^*	(4.8)−	Post-selection deme proportions
c	(4.8)	$(c_1, \ldots, c_K)^{\mathrm{T}} \in \operatorname{int} \Delta_K$
c	(4.119)	Scalar function of $x \in \bar{\Omega}$

Table 4.1. Continued

c^*	(4.8)	$(c_1^*, \ldots, c_K^*)^{\mathrm{T}} \in \operatorname{int} \Delta_K$
$\mathcal{D}$	(4.66)−	Subset of $\mathcal{J}$
d	(4.1a)−	Number of spatial dimensions
d_k	(4.60)	Column sums of $\operatorname{adj} V$
F	(4.52)	Logarithm of the geometric-mean fitness
f_i	(4.62)	Scalar functions in mapping $\Delta_J \to \Delta_J$
f	(4.130)	Scalar function of p in diallelic selection function
f^*	(4.133)+	$f^*(p) = f(1-p)$
G_i	(4.147)	Scalar function of $x \in \bar{\Omega}$
g	(4.68)	Product of gene frequencies
g	(4.129)	Spatial factor in diallelic selection function
H_i	(4.67)	Lyapunov function
H_i	(4.150)	Scalar function of $x \in \bar{\Omega}$
h	(4.2)+	Subscript for alleles
I	(4.120)−	$d \times d$ identity matrix
i	(4.1a)−	Subscript for alleles
J	(4.1a)−	Number of alleles
$\mathcal{J}$	(4.1a)	$\{1, \ldots, J\}$
$\mathcal{J}_i$	(4.1b)	$\{j \in \mathcal{J}: \ j \neq i\}$
$\mathcal{J}^{(k)}$	(4.41a)	Subset of $\mathcal{J}$
$\tilde{\mathcal{J}}$	(4.145)+	$\{2, \ldots, J-1\}$
j	(4.1b)	Subscript for alleles
K	(4.1a)−	Number of demes
$\mathcal{K}$	(4.1a)	$\{1, \ldots, K\}$
$\mathcal{K}_k$	(4.1b)	$\{\ell \in \mathcal{K}: \ \ell \neq k\}$
$\mathcal{K}^{(i)}$	(4.43)	Subset of $\mathcal{K}$
k	(4.1a)	Subscript for demes
k	(4.164)	Subscript for alleles
$\mathcal{L}$	(4.92)+	ω-limit set
L	(4.113a)	Migration operator
$L_1^{\dagger}$	(4.115)−	Adjoint of the closure of L and B
ℓ	(4.1b)	Subscript for demes
M	(4.7)−	Backward migration matrix
M	(4.107)−	Vector of drift coefficients
$\tilde{M}$	(4.9)−	Forward migration matrix
M_α	(4.107)−	Drift coefficients
$m_{k\ell}$	(4.6a)−	Elements of the backward migration matrix M
$\tilde{m}_{k\ell}$	(4.9)−	Elements of the forward migration matrix $\tilde{M}$
$m_{k\ell}^*$	(4.21)+	Scaled $m_{k\ell}$
m_k	(4.33)	Migration rates
n	(4.2)+	Subscript for demes
O	(4.28)	Order symbol
P_i	(4.22)	Mean frequency of A_i
P	(4.22)	$(P_1, \ldots, P_J)^{\mathrm{T}} \in \Delta_J$
$\hat{P}$	(4.44)−	Equilibrium value of $P \in \Delta_J$
$p_{i,k}$	(4.1b)+	Pre-selection frequency of A_i in deme k
p_i	(4.4a)	$(p_{i,1}, \ldots, p_{i,K})^{\mathrm{T}} \in [0,1]^K$
p_i	(4.47a)−	Frequency of A_i; in $[0,1]$

Table 4.1. Continued

Symbol	Reference	Definition
p_k	(4.92)−	Frequency of A_1 in deme k
p_i	(4.107)−	Frequency of A_i at (x,t)
$p^{(k)}$	(4.4b)	$(p_{1,k},\dots,p_{J,k})^{\mathrm{T}} \in \Delta_J$
p	(4.4c)	$(p^{(1)};\dots;p^{(K)})^{\mathrm{T}} \in \Delta_J^K$
p	(4.48)	$(p_1,\dots,p_J)^{\mathrm{T}} \in \Delta_J$
p	(4.94)+	$(p_1,p_2)^{\mathrm{T}} \in [0,1]^2$
p	(4.122)	Frequency of A_1; in $[0,1]$
$p'_{i,k}$	(4.6a)	$p_{i,k}(t+1)$
$p^*_{i,k}$	(4.6b)	Post-selection frequency of A_i in deme k
$\tilde{p}$	(4.42)+	Equilibrium value of $p \in \Delta_J^K$
$\tilde{p}$	(4.134b)+	Equilibrium value of $p \in [0,1]$
$\tilde{p}$	(4.178)−	Equilibrium value of $p \in \Delta_J$
$\tilde{p}_{i,k}$	(4.42)+	Components of $\tilde{p} \in \Delta_J^K$
$\tilde{p}^{(k)}$	(4.44)−	$(\tilde{p}_{1,k},\dots,\tilde{p}_{J,k})^{\mathrm{T}} \in \Delta_J$
$\hat{p}$	(4.31)+	Equilibrium value of $p \in \Delta_J^K$
$\hat{p}$	(4.61)+	$(\hat{p}_1,\dots,\hat{p}_J)^{\mathrm{T}} \in \Delta_J$
$\hat{p}$	(4.95a)−	Equilibrium value of $p \in [0,1]^2$
$\hat{p}$	(4.134b)+	Equilibrium value of $p \in [0,1]$
$\hat{p}_i$	(4.61)	Components of $\hat{p} \in \Delta_J$
p^*	(4.135)+	Equilibrium value of $p \in [0,1]$
p^{**}	(4.135)+	Equilibrium value of $p \in [0,1]$
$p^{(ij)}$	(4.162)−	Edge-equilibrium value of $p \in \Delta_J$
$p_h^{(ij)}$	(4.162)	Components of $p^{(ij)}$
$\tilde{p}_i$	(4.178)	Components of $\tilde{p} \in \Delta_J$
$q_{i,k}$	(4.23a)	$p_{i,k} - P_i$
q_i	(4.23b)	$p_i - P_i u \in \mathcal{R}^K$
q_i	(4.155a)	Frequency of A_i
$q^{(k)}$	(4.23c)	$p^{(k)} - P \in \mathcal{R}^J$
q	(4.23d)	$(q^{(1)};\dots;q^{(K)})^{\mathrm{T}} \in \mathcal{R}^{JK}$
q	(4.133)+	$1 - p \in [0,1]$
q	(4.154)	$(q_1,\dots,q_J)^{\mathrm{T}} \in \Delta_J$
q_i^*	(4.157a)	Frequency of A_i in (4.157)
q^*	(4.157a)	$(q_1^*,\dots,q_J^*)^{\mathrm{T}} \in \Delta_J$
$\hat{q}^*$	(4.157b)+	Equilibrium value of $q^* \in \Delta_J$
$\hat{q}$	(4.157b)+	Equilibrium value of $q \in \Delta_J$
$\mathcal{R}^J$	(4.3)	J-dimensional Euclidean space
$r_{ij,k}$	(4.20)	Selection coefficient of $A_i A_j$ in deme k
r_k	(4.32)	Selection coefficient of $A_1 A_1$ in deme k
r_i	(4.105)	Uniform selection coefficient of A_i
r_i	(4.109)−	Selection coefficient of A_i at (x,t)
$r_{i,k}$	(4.82b)	Selection coefficient of A_i in deme k
$\bar{r}_k$	(4.82b)	Mean selection coefficient in deme k
r_{ij}	(4.105)	Uniform selection coefficient of $A_i A_j$
r_{ij}	(4.109)−	Selection coefficient of $A_i A_j$ at $x \in \bar{\Omega}$
$\bar{r}$	(4.105)	Uniform mean selection coefficient
$\bar{r}$	(4.109)	Mean selection coefficient at (x,t)

Table 4.1. Continued

r^*	(4.141)	$\max_{j \in \mathcal{J}} r_j$
$\mathcal{S}_a$	(4.65)	$\{i \in \mathcal{J} : a_i > 0\}$
S_i	(4.110)	Selection function for A_i
S	(4.122)	Selection function for A_1
s_k	(4.32)	Selection coefficient of $A_2 A_2$ in deme k
s_i	(4.116)	Selection coefficient at $x \in \bar{\Omega}$ without dominance
$s_{i,k}$	(4.99)	Selection coefficient without dominance
$s_{i,k}^*$	(4.101)	$\max_{j \in \mathcal{J}_i} s_{j,k}$
$s_{i,k}^{**}$	(4.101)	$\min_{j \in \mathcal{J}_i} s_{j,k}$
$\bar{s}$	(4.117b)	Mean selection coefficient without dominance
s_i^*	(4.145)	$\max_{j \in \mathcal{J}_i} s_j$
$s^{(ij)}$	(4.163)	$\max_{h \in \mathcal{J}_i \cap \mathcal{J}_j} s_h$
$\tilde{s}^{(ij)}$	(4.163)	$\max[s_i, s_j]$
T_i	(4.154)−	Scalar function of x and p
$\bar{T}_i$	(4.156)	Average of T_i in Ω
t	(4.1b)+	Time in generations
$\tilde{t}$	(4.28)−	Characteristic equalization time
t_i	(4.148)+	Fixed time
u	(4.16)	$(1, \ldots, 1)^{\mathrm{T}} \in \mathcal{R}^K$
u	(4.113)	Scalar function of $x \in \bar{\Omega}$
V	(4.59)−	$J \times J$ matrix of viabilities
V	(4.107)−	$d \times d$ matrix of diffusion coefficients
$V_{\alpha\beta}$	(4.107)−	Diffusion coefficients
V_0	(4.120)−	Scalar function of $x \in \bar{\Omega}$
$v_{i,k}$	(4.11)	Viability without dominance
$\bar{v}_k$	(4.12b)	Mean viability without dominance
W	(4.111)−	Arbitrary $d \times d$ matrix
$W_{\alpha\beta}$	(4.111)	Elements of W
$w_{ij,k}$	(4.5)−	Viability of $A_i A_j$ in deme k
$w_{i,k}$	(4.5)	Viability of A_i in deme k
$\bar{w}_k$	(4.5)	Mean viability in deme k
w_{ij}	(4.31)+	Uniform viability of $A_i A_j$
$\tilde{w}$	(4.51)	Geometric-mean viability
x_k	(4.53)−	Viability of $A_1 A_1$ in deme k
x_α	(4.107)−	Components of $x \in \bar{\Omega}$
x^*	(4.54)	Harmonic-mean viability
x^*	(4.153)+	Fixed $x \in \bar{\Omega}$
$\bar{x}$	(4.55)	Arithmetic-mean viability
x	(4.57)	$(x_1, \ldots, x_K)^{\mathrm{T}} \in \mathcal{R}^K$
x	(4.107)−	$(x_1, \ldots, x_d)^{\mathrm{T}} \in \bar{\Omega}$
$x^{(i)}$	(4.147)+	Fixed $x \in \bar{\Omega}$
$x^{(ij)}$	(4.163)+	Fixed $x \in \bar{\Omega}$
$\tilde{x}$	(4.174b)−	Fixed $x \in \bar{\Omega}$
y_k	(4.53)−	Viability of $A_2 A_2$ in deme k
y^*	(4.53)	Harmonic-mean viability
$\bar{y}$	(4.57)−	Arithmetic-mean viability
y	(4.57)	$(y_1, \ldots, y_K)^{\mathrm{T}} \in \mathcal{R}^K$

Table 4.1. Continued

Symbol	Reference	Definition
z	(4.3)	Vector in Δ_J
z_i	(4.3)	Components of z
α	(4.21)	Maximal left eigenvector of M
α	(4.107)$-$	Subscript for components of $x \in \bar{\Omega}$
α	(4.171)	Positive constant
α_k	(4.21)	Components of $\alpha \in$ int Δ_K
β_i	(4.38)	Positive constants
β_i	(4.147)$+$	Integral
β	(4.107)$-$	Subscript for components of $x \in \bar{\Omega}$
Γ_i	(4.97a)	$\{j \in \mathcal{J}: \ \gamma_{ij} > 0\}$
Γ_i^*	(4.138a)	$\{i\} \cup \Gamma_i$
γ_ℓ	(4.46)$-$	Positive constants
γ_{ij}	(4.96a)	Nonnegative constants
γ	(4.171)$-$	Positive constant
Δ_J	(4.3)	Simplex in $\mathcal{R}^J$
Δ_J^K	(4.3)$+$	$(\Delta_J)^K$
int Δ_J	(4.8)	Interior of Δ_J
$\partial\Delta_J^K$	(4.13)$+$	Boundary of Δ_J^K
Δ	(4.19)$-$	Difference operator
$\hat{\Delta}_J$	(4.66)	Subset of Δ_J
$\tilde{\Delta}_J$	(4.67)$-$	$\hat{\Delta}_J(\{i\} \cup \mathcal{S}_b)$
$\Delta_J^{(i)}$	(4.136a)	$\{p \in \Delta_J: \ p_i = 0\}$
$\Delta_J^{(i,\delta)}$	(4.136b)	$\{p \in \Delta_J: \ p_j \geq \delta \ \forall j \neq i\}$
$\Delta_J^{(i)*}$	(4.138b)	$\{p \in \Delta_J: \ p_j > 0 \ \forall j \in \Gamma_i^*\}$
$\delta_{k\ell}$	(4.17)$+$	Kronecker delta
δ	(4.136a)$-$	Small positive constant
δ_i	(4.148)$+$	Small positive constant
ϵ	(4.17)	Small nonnegative parameter
ϵ_1	(4.164)$+$	Small positive constant
ϵ_2	(4.164)$+$	Small positive constant
ζ_i	(4.76)$-$	Positive constants
η_i	(4.77)	Nonnegative constants
θ	(4.21)$+$	Positive constant
θ_{ij}	(4.161)	Scalar edge equilibrium
κ_1	(4.28)$-$	Positive constant
κ	(4.34)	Migration–selection ratio
κ	(4.128)	Degree of dominance
κ^*	(4.36)	Migration–selection ratio
κ^{**}	(4.36)	Migration–selection parameter
λ_1	(4.28)$-$	Maximal nonunit eigenvalue of M
λ	(4.110)	Selection intensity
λ_0	(4.134b)$+$	Principal eigenvalue of (4.134)
$\tilde{\lambda}$	(4.135)$+$	Positive constant
λ_1^*	(4.160)	$\min_{i \in \mathcal{J}_1} \lambda_0(s_i - s_1)$

Table 4.1. Continued

λ_{ij}	(4.161)+	Nonnegative constant
$\tilde{\lambda}_{1h}$	(4.164)	$\min_{k \in \mathcal{J}_1 \cap \mathcal{J}_h} \lambda_{1k}$
$\mu_{k\ell}$	(4.17)	Backward migration rates
$\tilde{\mu}_{k\ell}$	(4.81)	Forward migration rates
μ_1	(4.93a)−	μ_{12}
μ_2	(4.93a)−	μ_{21}
μ	(4.154)	$1/\lambda$
ν_i	(4.77)	Positive constants
ν	(4.113b)+	Unit outward normal vector on $\partial\Omega$
Ξ_0	(4.27b)+	Set of equilibria of (4.27)
Ξ_ϵ	(4.27b)+	Set of equilibria of (4.6)
ξ_k	(4.77)	Positive constants
ξ	(4.173)	Positive number
$\pi_{i,k}$	(4.87)	Frequency of A_i in deme k
$\pi^{(k)}$	(4.89)	$(\pi_{1,k}, \ldots, \pi_{J,k})^{\mathrm{T}} \in \Delta_J$
ρ_{ij}	(4.25a)	Mean selection coefficient of $A_i A_j$
ρ_i	(4.25b)	Mean selection coefficient of A_i
$\bar{\rho}$	(4.25c)	Mean selection coefficient
ρ	(4.107)−	Population density
Σ_0	(4.18)+	Set of equilibria of (4.18)
Σ_ϵ	(4.18)+	Set of equilibria of (4.6)
σ_k	(4.94)	μ_k/s_k
σ_i	(4.144)	Spatially averaged selection coefficient
σ_i	(4.158)	Spatially averaged selection coefficient
τ	(4.26)	Scaled time
τ	(4.154)	Scaled time
ϕ	(4.79b)	Positive constant
ϕ	(4.134)	Scalar eigenfunction
ϕ_k	(4.92)	Scalar selection functions
χ	(4.19)	Scalar fitness function
ψ	(4.79b)	Nonnegative constant
ψ	(4.115)−	Normalized principal eigenfunction of $L_1^\dagger$
Ω_0	(4.35)−	Region in Fig. 4.2
Ω_1	(4.35)−	Region in Fig. 4.2
Ω_+	(4.35)	Region in Fig. 4.2
Ω_-	(4.37)	Hatched region in Fig. 4.3
Ω	(4.107)−	Bounded, open domain in $\mathcal{R}^d$
$\bar{\Omega}$	(4.107)−	$\Omega \cup \partial\Omega$
$\partial\Omega$	(4.113b)+	Boundary of habitat Ω
Ω_i	(4.151)	Subset of Ω
ω_{ij}	(4.29)	Approximate mean viability of $A_i A_j$
ω_i	(4.29)	Approximate mean viability of A_i
$\bar{\omega}$	(4.29)	Approximate mean viability

References

Akin, E., Hofbauer, J., 1982. Recurrence of the unfit. Math. Biosci. 61, 51–62.

Akin, E., Losert, V., 1984. Evolutionary dynamics of zero-sum games. J. Math. Biol. 20, 231–258.

Atkinson, F. V., Watterson, G. A., Moran, P. A. P., 1960. A matrix inequality. Quart. J. Math. 11, 137–140.

Avise, J. C., 1994. Molecular Markers, Natural History and Evolution. Chapman and Hall, New York.

Barton, N. H., Shpak, M., 2000. The effect of epistasis on the structure of hybrid zones. Genet. Res. 75, 179–198.

Bulmer, M. G., 1972. Multiple niche polymorphism. Am. Nat. 106, 254–257.

Bürger, R., 2000. The Mathematical Theory of Selection, Recombination, and Mutation. Wiley, Chichester, UK.

Cannings, C., 1971. Natural selection at a multiallelic autosomal locus with multiple niches. J. Genet. 60, 255–259.

Carvalho, A. N., Hale, J. K., 1991. Large diffusion with dispersion. Nonlin. Anal. 17, 1139–1151.

Conway, E., Hoff, D., Smoller, J., 1978. Large time behavior of solutions of systems of nonlinear reaction-diffusion equations. SIAM J. Appl. Math. 35, 1–16.

ten Eikelder, H. M. M., 1979. A nonlinear diffusion problem arising in population genetics. Report NA-25, Delft University of Technology.

Endler, J. A., 1977. Geographical Variation, Speciation, and Clines. Princeton University Press, Princeton, NJ.

Eyland, E. A., 1971. Moran's island model. Genetics 69, 399–403.

Fife, P. C., Peletier, L. A., 1981. Clines induced by variable selection and migration. Proc. Roy. Soc. London B 214, 99–123.

Fleming, W. H., 1975. A selection-migration model in population genetics. J. Math. Biol. 2, 219–233.

Gantmacher, F. R., 1959. The Theory of Matrices, vol. II. Chelsea, New York.

Graybill, F. A., 1969. Introduction to Matrices, with Applications in Statistics. Wadsworth, Belmont, CA.

Hadeler, K. P., 1981. Diffusion in Fisher's population model. Rocky Mtn. J. Math. 11, 39–45.

Hadeler, K. P., Glas, D., 1983. Quasimonotone systems and convergence to equilibrium in a population genetic model. J. Math. Anal. Appl. 95, 297–303.

Haldane, J. B. S., 1948. The theory of a cline. J. Genet. 48, 277–284.

Henry, D., 1981. Geometric Theory of Semilinear Parabolic Equations. Lecture Notes in Mathematics, vol. 840. Springer, Berlin.

Hirsch, M. W., 1982. Systems of differential equations which are competitive or cooperative. I: Limit sets. SIAM J. Math. Anal. 13, 167–179.

Hofbauer, J., Sigmund, K., 1988. The Theory of Evolution and Dynamical Systems. Cambridge University Press, Cambridge.

Karlin, S., 1977. Gene frequency patterns in the Levene subdivided population model. Theor. Popul. Biol. 11, 356–385.

Karlin, S., 1978. Theoretical aspects of multi-locus selection balance. Pp. 503–587 in: Levin, S. A. (Ed.), Studies in Mathematical Biology. Studies in Mathematics, vol. 16. Mathematical Association of America, Washington.

Karlin, S., 1982. Classification of selection-migration structures and conditions for a protected polymorphism. Evol. Biol. 14, 61–204.

Karlin, S., 1984. Mathematical models, problems, and controversies of evolutionary theory. Bull. Am. Math. Soc. 10, 221–273.

Karlin, S., McGregor, J., 1972a. Application of method of small parameters to multi-niche population genetic models. Theor. Popul. Biol. 3, 186–209.

Karlin, S., McGregor, J., 1972b. Polymorphisms for genetic and ecological systems with weak coupling. Theor. Popul. Biol. 3, 210–238.

Kingman, J. F. C., 1961. On an inequality in partial averages. Quart. J. Math. 12, 78–80.

Levene, H., 1953. Genetic equilibrium when more than one ecological niche is available. Am. Nat. 87, 311–313.

Li, C. C., 1955. The stability of an equilibrium and the average fitness of a population. Am. Nat. 89, 281–295.

Losert, V., Akin, E., 1983. Dynamics of games and genes: discrete versus continuous time. J. Math. Biol. 17, 241–251.

Lou, Y., Nagylaki, T., 2002. A semilinear parabolic system for migration and selection in population genetics. J. Diff. Eqs. 181, 388–418.

Lou, Y., Nagylaki, T., 2004. Evolution of a semilinear parabolic system for migration and selection in population genetics. J. Diff. Eqs. 204, 292–322.

Lou, Y., Nagylaki, T., 2006. Evolution of a semilinear parabolic system for migration and selection without dominance. J. Diff. Eqs. 225, 624–665.

Lui, R., 1986. A nonlinear integral operator arising from a model in population genetics. IV. Clines. SIAM J. Math. Anal. 17, 152–168.

Lyubich, Yu. I., 1992. Mathematical Structures in Population Genetics. Biomathematics, vol. 22. Springer, Berlin.

Lyubich, Yu. I., Maistrovskii, G. D., Ol'khovskii, Yu. G., 1980. Selection-induced convergence to equilibrium in a single-locus autosomal population. Probl. Inf. Transm. 16, 66–75.

Maynard Smith, J., 1970. Genetic polymorphism in a varied environment. Am. Nat. 104, 487–490.

Moody, M. E., 1979. Polymorphism with migration and selection. J. Math. Biol. 8, 73–109.

Mulholland, H. P., Smith, C. A. B., 1959. An inequality arising in genetical theory. Am. Math. Mon. 66, 673–683.

Nagylaki, T., 1975. Conditions for the existence of clines. Genetics 80, 595–615.

Nagylaki, T., 1976. Clines with variable migration. Genetics 83, 867–886.

Nagylaki, T., 1978a. Random genetic drift in a cline. Proc. Natl. Acad. Sci. USA 75, 423–426.

Nagylaki, T., 1978b. A diffusion model for geographically structured populations. J. Math. Biol. 6, 375–382.

Nagylaki, T., 1978c. Clines with asymmetric migration. Genetics 88, 813–827.

Nagylaki, T., 1979. Migration-selection polymorphisms in dioecious populations. J. Math. Biol. 8, 123–131.

Nagylaki, T., 1980. The strong-migration limit in geographically structured populations. J. Math. Biol. 9, 101–114.

Nagylaki, T., 1983. Evolution of a large population under gene conversion. Proc. Natl. Acad. Sci. USA 80, 5941–5945.

Nagylaki, T., 1989. The diffusion model for migration and selection. Pp. 55–75 in: Hastings, A. (Ed.), Some Mathematical Questions in Biology. Lecture Notes on Mathematics in the Life Sciences, vol. 20. American Mathematical Society, Providence, RI.

Nagylaki, T., 1992. Introduction to Theoretical Population Genetics. Biomathematics, vol. 21. Springer, Berlin.

Nagylaki, T., 1993. The evolution of multilocus systems under weak selection. Genetics 134, 627–647.

Nagylaki, T., 1996. The diffusion model for migration and selection in a dioecious population. J. Math. Biol. 34, 334–360.

Nagylaki, T., 1997. The diffusion model for migration and selection in a plant population. J. Math. Biol. 35, 409–431.

Nagylaki, T., 1998. The expected number of heterozygous sites in a subdivided population. Genetics 149, 1599–1604.

Nagylaki, T., Hofbauer, J., Brunovský, P., 1999. Convergence of multilocus systems under weak epistasis or weak selection. J. Math. Biol. 38, 103–133.

Nagylaki, T., Lou, Y., 2001. Patterns of multiallelic polymorphism maintained by migration and selection. Theor. Popul. Biol. 59, 297–313.

Nagylaki, T., Lou, Y., 2006a. Multiallelic selection polymorphism. Theor. Popul. Biol. 69, 217–229.

Nagylaki, T., Lou, Y., 2006b. Evolution under the multiallelic Levene model. Theor. Popul. Biol. 70, 401–411.

Nagylaki, T., Lou, Y., 2007a. Evolution under multiallelic selection-migration models. Theor. Popul. Biol. 72, 21–40.

Nagylaki, T., Lou, Y., 2007b. Evolution at a multiallelic locus under migration and uniform selection. J. Math. Biol. 54, 787–796.

Nagylaki, T., Lucier, B., 1980. Numerical analysis of random drift in a cline. Genetics 94, 497–517.

Pauwelussen, J. P., 1981. Nerve impulse propagation in a branching nerve system: A simple model. Physica D 4, 67–88.

Pauwelussen, J. P., Peletier, L. A., 1981. Clines in the presence of asymmetric migration. J. Math. Biol. 11, 207–233.

Protter, M. H., Weinberger, H. F., 1984. Maximum Principles in Differential Equations. 2nd ed. Springer, Berlin.

Prout, T., 1968. Sufficient conditions for multiple niche polymorphism. Am. Nat. 102, 493–496.

Redlinger, R., 1983. Über die C^2-Kompaktheit der Bahn der Lösungen semilinearer parabolischer Systeme. Proc. Roy. Soc. Edinb. A 93, 99–103.

Scheuer, P. A. G., Mandel, S. P. H., 1959. An inequality in population genetics. Heredity 13, 519–524.

Senn, S., Hess, P., 1982. On positive solutions of a linear elliptic eigenvalue problem with Neumann boundary conditions. Math. Ann. 258, 459–470.

Some Challenging Mathematical Problems in Evolution of Dispersal and Population Dynamics

Y. Lou

Department of Mathematics,
The Ohio State University, Columbus, OH 43210, USA
email: lou@math.ohio-state.edu

Summary. We discuss the effects of dispersal (either random or biased) and spatial heterogeneity on population dynamics via reaction–advection–diffusion models. We address the question of determining optimal spatial arrangement of resources and study how advection along resource gradients affects the extinction of species. The effects of dispersal and spatial heterogeneity on the total population size of single species are carefully investigated, along with some other properties of species. These properties have important applications to invasions of rare species. Some interesting connection between the evolution of unconditional dispersal and diffusion-driven extinction is revealed. We also investigate the outcome of competition for two similar species, and show how invasion and coexistence are affected by resource utilization, inter-specific competition, and dynamics of habitat edges. In particular, interesting effects of intermediate values of dispersal rates are found. The evolution of conditional dispersal is also addressed, and we illustrate that the geometry of a habitat can play an important role in the evolution of conditional dispersal and that strong directed movement of species can induce the coexistence of competing species. If both species disperse by random diffusion and advection along environmental gradients and one species has much stronger biased movement than the other one, then at least two scenarios can occur: either both species can coexist or the "smarter" species is always the loser. These results seem to suggest that selection is against large advection along resource gradient and that an intermediate biased movement rate may evolve. Numerous open problems will be discussed.

5.1 Introduction

The evolution of dispersal has been one of the central topics in recent theoretical studies of population dynamics. Metapopulation models have been widely used for spatially structured populations (Lehman and Tilman [78]; Hanski [52]), and one of the primary criticisms for these models is that the dispersal functions are usually so simplified that the results can be misleading (Travis and French [120]). In fact, classical metapopulation models often are not even spatially explicit and thus it is difficult to use them to gain insight into the effects of different dispersal mechanisms. Models such as those used in [61, 94] are more accurately described as discrete diffusion models and they do incorporate mechanistic assumptions about dispersal and treat population dynamics explicitly. We refer to Sects. 1.4 and 2.6.2 of [12] for more detailed discussions.

Two theories dominate current investigations on dispersal: (1) the "source–sink model" (Hastings [54]; Holt [60]; Hanski and Thomas [53]; Doebeli [36]; Travis and Dytham [119]), in which individuals disperse at fixed constant rates, regardless of the local environment; and (2) the "balanced dispersal model" (McPeek and Holt [94]), in which the dispersal is conditional since the dispersal rate depends on a combination of local biotic and abiotic factors such as habitat quality. It is well accepted that conditional dispersal can be a crucial factor in population dynamics (Turchin [121]; Travis and French [120]; Bowler and Benton [6]; Armsworth and Roughgarden [1, 2]).

One central question is which patterns of dispersal can confer some sort of selective or ecological advantage? When habitat quality varies spatially but remains constant in time, the source–sink model predicts that the selection is for slow dispersal (Hastings [54]; Holt [60]). However, McPeek and Holt [94] showed that when there is conditional dispersal, e.g., with dispersal depending on patch carrying capacity, dispersal can evolve in a spatially varying environment.

In reality, species are neither completely ignorant of the surrounding environment nor will their movement perfectly track resource gradients. It is more likely that their movements combine both random and directed ones: for instance, adding some amount of random movement to a purely directed movement strategy might help an individual escape a local trap to find distant but better sources (Armsworth and Roughgarden [1]). In other words, a balanced combination of both random and directed movement might help the species better utilize available resources and thus maximize its chance of survival.

Since the pioneering work of Skellam [117], there have been tremendous theoretical advances in reaction–diffusion modeling of invasions and species interactions (Okubo and Levin [106]; Murray [101]; Shigesada and Kawasaki [115]; Cantrell and Cosner [12]). The book by Cantrell and Cosner [12], which covers many major issues in spatial ecology via the reaction–diffusion approach, motivates this current survey. We will discuss some of the most recent progress on applications of reaction–advection–diffusion equations to population dynamics and the evolution of dispersal, including both unconditional and conditional dispersal. Our main goal is to introduce the readers to the current state of the art in this area and raise some questions for further research.

This paper is organized as follows. In Sect. 5.2 we discuss the question of determining the optimal spatial arrangement of resources and study the effects of advection along resource gradients on extinction of species.

Section 5.3 is devoted to studying single species, with the main focus on the effects of dispersal and spatial heterogeneity on the total population size and other properties of species. These properties have important implications for invasions of rare species.

Sections 5.4–5.6 are devoted to the study of two competing species in a heterogeneous environment. In Sect. 5.4 we investigate the evolution of unconditional dispersal, diffusion-driven extinction, and their intimate connections.

Section 5.5 is concerned with the outcome of competition for two similar species, and we show how invasion and coexistence are affected by resource utilization, interspecific competition, and dynamics of habitat edges. In particular, interesting effects of *intermediate* values of dispersal rates are found.

Finally in Sect. 5.6 we study the evolution of conditional dispersal and show how the geometry of a habitat can play an important role in the evolution of dispersal, and that strong directed movement of species can induce the coexistence of competing species. We also consider the situation when both species disperse by random diffusion and advection along environmental gradients and one species has much stronger biased movement than the other one. The species with stronger biased movement behaves like a specialist as it mainly pursues resources at places of locally most favorable environments. The other species has a rather balanced mixed dispersal strategy and can be regarded as a generalist. It is shown that at least two scenarios can occur: If the generalist's biased movement rate is relatively smaller than its own random movement rate, then both species can coexist; If its biased movement is relatively stronger than its random movement, then the generalist is always the winner, regardless of the initial condition. These results seem to suggest that selection is against large advection and that an intermediate biased movement rate may evolve.

Throughout this survey numerous open problems will be discussed and some conjectures will be presented. We also refer to recent excellent surveys of Ni [103, 104] on competition models with density-dependent diffusion, Levin et al. [79] on the evolution of plant dispersal, Neuhauser [102] on dynamics of metapopulation models.

5.2 Linear Eigenvalue Problems with Indefinite Weight

In this section we discuss two linear eigenvalue problems with indefinite weight subject to no-flux boundary conditions. These linear eigenvalue problems are particularly important since they are closely connected to predictions of persistence vs. extinction of species in reaction–diffusion models.

Section 5.2.1 is concerned with the question of determining the optimal spatial arrangement of favorable and unfavorable regions for species to survive, and the conclusion for a one-dimensional closed habitat is that a single favorable region at one of the two ends of the whole habitat provides the best opportunity for the species to survive.

In Sect. 5.2.2 we study the effects of advection along environmental gradients on dynamics of logistic type reaction–diffusion models for population growth. The local population growth rate is assumed to be spatially inhomogeneous, and the advection is taken to be a multiple of the gradient of the local population growth rate. It is also assumed that the boundary acts as a reflecting barrier to the population. It turns out that the effects of such advection depends crucially on the shape of the habitat: if the habitat is convex, the movement in the direction of the gradient of the growth rate is always beneficial to the population, while such advection can be harmful for certain nonconvex habitats.

These results depend crucially on boundary conditions, e.g., the results for Dirichlet are different [4, 8, 9].

5.2.1 Rearrangement of Resources

The linear eigenvalue problem with indefinite weight

$$
\begin{cases}
\Delta\varphi + \lambda m(x)\varphi = 0 & \text{in } \Omega, \\
\dfrac{\partial\varphi}{\partial n} = 0 & \text{on } \partial\Omega
\end{cases}
\tag{5.1}
$$

and its variants have been extensively investigated for last two decades since they play crucial roles in studying nonlinear mathematical models from population biology. Throughout this paper we shall assume that the habitat Ω is a bounded region in $\mathbb{R}^N$ with smooth boundary $\partial\Omega$ and n is the outward unit normal vector on $\partial\Omega$. The zero-flux boundary condition in (5.1) means that no individuals cross the boundary of the habitat.

The function $m(x)$ represents local intrinsic growth rate of species at location x. It reflects the quality and quantity of resources available at the point x, and is often referred as an indefinite weight since it may change sign in Ω. For this subsection, we assume that the function m is nonconstant, bounded, and measurable in Ω. Define

$$
\Omega_+ = \{x \in \Omega : m(x) > 0\}, \qquad \Omega_- = \{x \in \Omega : m(x) < 0\}.
$$

The subdomain Ω_+ can be viewed as a *source* as the species has positive intrinsic growth rate there; likewise, the region Ω_- can be regarded as a *sink*.

We call λ a *principal eigenvalue* of (5.1) if λ has a positive eigenfunction $\varphi \in H^1(\Omega)$. By elliptic regularity and the Sobolev embedding theorem [47], the function φ satisfies $\varphi \in W^{2,q}(\Omega) \cap C^{1,\gamma}(\overline{\Omega})$ for every $q > 1$ and every $\gamma \in (0,1)$, and $\varphi > 0$ in $\overline{\Omega}$. Clearly, $\lambda = 0$ is a principal eigenvalue of (5.1) with positive constants as its eigenfunctions. Of particular importance and interest is the existence of *positive* principal eigenvalues.

If (5.1) has a positive eigenvalue $\lambda_1(m)$ with corresponding positive eigenfunction φ_1, integrating the equation of φ_1 we have

$$
\int_\Omega m\varphi_1 = 0,
$$

which implies that both Ω_+ and Ω_- have positive Lebesgue measure; dividing the equation of φ_1 by φ_1 and then integrating in Ω, we find

$$\lambda_1(m) \int_\Omega m = - \int_\Omega \frac{|\nabla\varphi_1|^2}{\varphi_1^2} < 0$$

since φ_1 is not equal to any positive constant (as m is not identically equal to any constant). In summary, the condition

(A1) The set Ω_+ has positive Lebesgue measure, and $\int_\Omega m < 0$

is necessary for the existence of a positive principal eigenvalue. This condition turns out to be also sufficient as shown by the following result [4, 7, 44, 56, 114]:

Theorem 5.2.1 *The eigenvalue problem* (5.1) *has a positive principal eigenvalue (denoted by $\lambda_1(m)$) if and only if* (A1) *holds. Moreover, $\lambda_1(m)$ is the only positive principal eigenvalue and it is simple; it is also the smallest positive eigenvalue of* (5.1), *and is given by*

$$\lambda_1(m) = \inf_{\varphi \in S(m)} \frac{\int_\Omega |\nabla\varphi|^2}{\int_\Omega m(x)\varphi^2}, \tag{5.2}$$

where

$$S(m) := \left\{ \varphi \in H^1(\Omega) : \int_\Omega m(x)\varphi^2 > 0 \right\}.$$

As an application, we consider the diffusive logistic equation

$$\begin{aligned}
u_t &= \Delta u + \lambda u[m(x) - u] \quad &&\text{in } \Omega \times \mathbb{R}^+, \\
\frac{\partial u}{\partial n} &= 0 \quad &&\text{on } \partial\Omega \times \mathbb{R}^+, \\
u(x,0) &\geq 0, \quad u(x,0) \not\equiv 0 \quad &&\text{in } \overline{\Omega},
\end{aligned} \tag{5.3}$$

where $u(x,t)$ represents the density of a species at location x and time t. Hence, only nonnegative solutions of (5.3) are of interest. Again, the function $m(x)$ represents the intrinsic growth rate of a species, which is positive in the favorable part of habitat (Ω_+) and negative in unfavorable one (Ω_-). The integral $\int_\Omega m$ can be viewed as a measure of the total resources in a spatially heterogeneous environment.

So far as invasion of species is concerned, λ^{-1} acts like the diffusion coefficient. To see this, we set $\mu = \lambda^{-1}$ and introduce the new time variable τ with $\tau = \lambda t$. Then the equation of u can be rewritten as

$$u_\tau = \mu\Delta u + u[m(x) - u].$$

Set $\mu_1(m) = \lambda_1^{-1}(m)$ (we define $\mu_1(m) = +\infty$ if $\lambda_1(m) = 0$). It is well-known that:

(1) If $\mu \geq \mu_1(m)$, i.e., $\lambda \leq \lambda_1(m)$, then $u(x,t) \to 0$ uniformly in $\overline{\Omega}$ as $t \to \infty$ for all nonnegative and nontrivial initial data, i.e., the species goes to extinction.

(2) If $\mu < \mu_1(m)$, i.e., $\lambda > \lambda_1(m)$, then $u(x,t) \to u^*(x)$ uniformly in $\overline{\Omega}$ as $t \to \infty$, where u^* is the unique positive steady state of (5.3) in $W^{2,q}(\Omega)$ for every $q > 1$, i.e., the invasion occurs.

We are mainly concerned with the dependence of the principal eigenvalue $\lambda_1(m)$ of (5.1) on the weight function $m(x)$. In particular, we are interested in how spatial variation in the environment of the habitat affects the maintenance of species. To be more precise, let m_0 be some constant satisfying $m_0 \in (0,1)$, and assume

(A2) $-1 \le m(x) \le 1$ a.e. in Ω, and $\int_\Omega m \le -m_0|\Omega|$.

We address the following mathematical question:

Question. Among all functions $m(x)$ that satisfy (A1) and (A2), which $m(x)$ will yield the smallest $\lambda_1(m)$?

This question is motivated by the following biological consideration: the species can survive if and only if $\lambda > \lambda_1(m)$, and the smaller $\lambda_1(m)$ is, the more likely that the species can exist. Biologically, it is equivalent to determining the optimal spatial arrangement of the favorable and unfavorable parts of the habitat for species to survive. This question was first addressed by Cantrell and Cosner in [8, 9], and it remains largely open.

The question is also mathematically meaningful: by (A1) and Theorem 5.2.1, the positive principal eigenvalue $\lambda_1(m)$ exists and is unique. Furthermore, the infimum of $\lambda_1(m)$ among all those m that satisfy both (A1) and (A2) is positive by a result of Saut and Scheurer [113]:

Theorem 5.2.2 *Suppose that* (A1) *holds. Then*

$$\lambda_1(m) \ge \frac{\nu_1 \left| \int_\Omega m \right|}{\int_\Omega m^2(x)\,dx + \left| \int_\Omega m \right| \sup_\Omega m},$$

where ν_1 is the smallest positive eigenvalue of the Laplace operator with homogeneous zero Neumann boundary condition.

For any $m_0 \in (0,1)$, we define

$$\mathcal{M} = \{m \in L^\infty(\Omega) : m(x) \text{ satisfies (A1) and (A2)}\}, \qquad (5.4)$$

and set

$$\lambda_{\text{inf}} := \inf_{m \in \mathcal{M}} \lambda_1(m).$$

As an immediate consequence of Theorem 5.2.2, we have

$$\lambda_{\text{inf}} \ge \frac{\nu_1 m_0}{1 + m_0} > 0.$$

The existence and profile of global minimizers of $\lambda_1(m)$ in $\mathcal{M}$ with Dirichlet boundary condition was first addressed by Cantrell and Cosner in [8], in which

Cantrell and Cosner showed that there exists some measurable set $E \subset \Omega$ with $|E| > 0$ such that $\lambda_1(\chi_E - \chi_{\Omega \backslash E}) = \lambda_{\inf}$. The result of Cantrell and Cosner can be viewed as saying that there exists a "bang–bang" type optimal control for minimizing $\lambda_1(m)$ in $\mathcal{M}$. For Neumann boundary condition, a stronger result is established in [90]:

Theorem 5.2.3 *The infimum λ_{inf} is attained by some $m \in \mathcal{M}$. Moreover, if $\lambda_1(m) = \lambda_{inf}$, then m can be represented as $m(x) = \chi_E - \chi_{\Omega \backslash E}$ a.e. in Ω for some measurable set $E \subset \Omega$.*

Theorem 5.2.3 implies that the global minimizers of $\lambda_1(m)$ in $\mathcal{M}$ must be of "bang–bang" type, and is indeed contained in the set

$$\mathcal{M}_\alpha = \left\{ m : m = \chi_E - \chi_{\Omega \backslash E} \text{ for some } E \subset \Omega \text{ with } |E| = \alpha|\Omega| \right\},$$

where $\alpha = (1 - m_0)/2$. It is easy to check that every $m \in \mathcal{M}_\alpha$ satisfies $\int_\Omega m = -m_0|\Omega|$, i.e., $\mathcal{M}_\alpha \subset \mathcal{M}$.

By Theorem 5.2.3 and the above discussion, in order to determine all of the global minimizers of $\lambda_1(m)$ in $\mathcal{M}$, it suffices to characterize $E \subset \Omega$ such that the corresponding weight function $m(x) = \chi_E - \chi_{\Omega \backslash E}$ minimizes the principal eigenvalue $\lambda_1(m)$ in $\mathcal{M}_\alpha$. Recently, the following complete characterization of all global minimizers of $\lambda_1(m)$ in $\mathcal{M}$ when $N = 1$ are established in [90]:

Theorem 5.2.4 *Suppose that $N = 1$, $\Omega = (0,1)$, and $\alpha = (1 - m_0)/2$. Then $\lambda_1(m) = \lambda_{inf}$ for some function $m \in \mathcal{M}$ if and only if $m = \chi_E - \chi_{\Omega \backslash E}$ a.e. in $(0,1)$, where either $|E \cap (0,\alpha)| = \alpha$ or $|E \cap (1-\alpha,1)| = \alpha$.*

Theorem 5.2.4 implies that when Ω is an interval, then there are exactly two global minimizers of $\lambda_1(m)$ (up to change of a set of measure zero). This substantially improves previous work in the one-dimensional case. Cantrell and Cosner in [9] studied the case when Ω is the unit interval $(0,1)$ under three different boundary conditions (Dirichlet, Neumann, and Robin type). For Neumann boundary condition, they showed that if $m(x) \equiv 1$ on a "single" subinterval of fixed length and $m(x) = -1$ on the remainder of the interval, then the smallest value of $\lambda_1(m)$ with $m(x)$ so restricted occurs when the subinterval where $m(x) \equiv 1$ is at one of the ends of the interval $(0,1)$; they also considered the situation when $m(x) \equiv -1$ in a "single" subinterval of fixed length, and $m(x) \equiv 1$ on the remainder of $(0,1)$. They proved that in the latter case, the smallest value for $\lambda_1(m)$ occurs when $m(x) \equiv -1$ at one of the ends of $(0,1)$. Some new ideas based on a characterization of critical points and continuous dependence of $\lambda_1(m)$ with respect to m are given in [90], and the analysis in [90] can also be useful in handling Dirichlet, Robin, and periodic boundary conditions.

The characterization of the optimal set E in $\mathcal{M}_\alpha$ for higher-dimensional domains is a difficult problem, even for two-dimensional balls or rectangles (see [73] for some recent progress on cylindrical domains). For general domains, we have the following

Conjecture. If Ω is convex, then both E and Ω/E are connected, and $\partial E \cap \partial \Omega$ has positive surface measure.

For the connectivity of E, the geometry of Ω might be irrelevant. However, the convexity assumption on Ω appears necessary in order to ensure the connectivity of Ω/E since for dumbbell type domains, Ω/E may have multiple connected components. We also conjecture that if Ω is a ball, then E is rotationally symmetric with respect to some diameter of the ball.

Similarly, little is known for eigenvalue problem (5.1) but with Dirichlet, Robin and periodic boundary conditions in higher dimensions. We refer to [5] for recent progress in related problems, where symmetrization methods are extensively employed. The symmetrization idea for the Dirichlet case was also discussed earlier in [8]. For other related works, see [42, 76, 77] and references therein.

5.2.2 Biased Movement of Species

If the environment is spatially heterogeneous, the population may have a tendency to move along the gradient of the resources in addition to random dispersal. This leads Belgacem and Cosner [4] to add an advection term to (5.3) and consider the model

$$
\begin{cases}
\dfrac{\partial u}{\partial t} = \nabla \cdot [\nabla u - \alpha u \nabla m] + \lambda (m - u)u & \text{in } \Omega \times (0, \infty), \\[2mm]
\dfrac{\partial u}{\partial n} - \alpha u \dfrac{\partial m}{\partial n} = 0 & \text{on } \partial \Omega \times (0, \infty).
\end{cases}
\tag{5.5}
$$

The flux of population density is given by $\mathbf{J} = -\nabla u + \alpha u \nabla m$, where α measures the tendency of the population to move up along the gradient of $m(x)$, and $m(x)$ is assumed to be twicely continuously differentiable in $\bar{\Omega}$.

It is also shown in [4] that the effects of the advection term $\alpha u \nabla m$ depend critically on boundary conditions: for no-flux boundary conditions, sufficiently rapid movement in the direction of $m(x)$ is always beneficial, and the movement up the gradient of $m(x)$ may be either beneficial or harmful for zero Dirichlet boundary condition.

According to [4], for every α, there exists an unique nonnegative constant $\lambda_* = \lambda_*(\alpha)$ such that the following holds:

(1) If $\lambda > \lambda_*$, (5.5) has a unique positive steady state, which is globally attracting among nontrivial nonnegative solutions of (5.5);
(2) If $\lambda_* > 0$ and $0 < \lambda \le \lambda_*$, then all nonnegative solutions of (5.5) converge to zero as $t \to \infty$.

The constant λ_* is the principal eigenvalue of an eigenvalue problem related to (5.5). By [4] (see also [3]), λ_* can be characterized by

$$
\lambda_* := \inf_{\varphi \in S} \frac{\int_\Omega e^{\alpha m} |\nabla \varphi|^2}{\int_\Omega e^{\alpha m} m \varphi^2},
\tag{5.6}
$$

where

$$S = \left\{ \varphi \in W^{1,2}(\Omega) : \int_{\Omega} e^{\alpha m} m \varphi^2 > 0 \right\}. \tag{5.7}$$

It can be shown [31] that if $\int_{\Omega} m \geq 0$, then $\lambda_* \equiv 0$ for $\alpha \geq 0$; if m changes sign and $\int_{\Omega} m < 0$, then there exists a unique $\alpha_* > 0$ such that $\lambda_* > 0$ if $\alpha < \alpha_*$, and $\lambda_* \equiv 0$ if $\alpha \geq \alpha_*$. Throughout this subsection, we shall also assume (A1).

Question. If we start with $\alpha = 0$, i.e., without directed motion up the gradient of $m(x)$, is increasing α always beneficial to the survival of population?

In terms of λ_*, for small α, this is equivalent to asking whether $\lambda_*'(0) < 0$. Since the population can survive if and only if $\lambda > \lambda_*$, intuitively we may expect that the smaller λ_* is, the more likely it is that the population will survive. In this connection, the following result is proved in [31].

Theorem 5.2.5 *Suppose that* (A1) *holds.*
(i) *For any convex domain Ω and any $m(x)$, $\lambda_*'(0) < 0$;*
(ii) *There exist some nonconvex Ω and $m(x)$ such that $\lambda_*'(0) > 0$.*

Part (ii) seems to be counterintuitive because movement in the increasing direction of the local growth rate should always be helpful to the population. One possible explanation is that for some nonconvex habitats such as dumbbell type ones, an individual moving in the direction of increasing $m(x)$ may "hit" the boundary of the habitat and cannot go further, but a randomly moving individual moves in all possible directions with equal probability, and thus might eventually be able to "turn the corner" and move into the region with more favorable resource levels.

The type of domains constructed in (ii) are so called thin domains, and there has been some active research on the equilibria and dynamics of evolution equations in thin domains: e.g., bistable scalar equation in thin tubular domains [123], Navier–Stokes equations in thin 3D domains [112], reaction–diffusion equations in thin domains [50], the Lotka–Volterra competition–diffusion system in thin tubular domains [72], Ginzburg–Landau equation in thin domains [25], subharmonic solutions [108], etc.; See also the survey [111] about PDEs in thin domains.

Theorem 5.2.5 is a local result that deals only with small α. With some extra conditions on $m(x)$, the following global result is established in [31].

Theorem 5.2.6 *Suppose that* (A1) *holds, Ω is convex, and the matrix $(m_{x_i x_j})_{1 \leq i,j \leq n}$ is nonpositive. Then $\lambda_*'(\alpha) < 0$ for all $0 \leq \alpha < \alpha_*$. In particular, this implies that*

$$\lambda_*(0) = \max_{\alpha \geq 0} \lambda_*(\alpha). \tag{5.8}$$

The biological meaning of (5.8) is that the movement in the increasing direction of m is always beneficial to the population. For general $m(x)$, we have the following

Conjecture. If Ω is convex, then (5.8) always holds.

5.3 Logistic Models for Single Species

The dynamics of mathematical models for single species are not only of independent interest, they are also quite crucial in studying dynamics of multiple interacting species, especially issues concerning invasions of rare species. In this section we focus on logistic type population models with either random diffusion or biased movement along the gradient of resources. In Sect. 5.3.1 we study the effects of dispersal rate and spatial heterogeneity on the total population size. Section 5.3.2 is concerned with how biased movement of species affects the total population size and the density distribution of species.

5.3.1 Total Population Size

In this subsection we study the effects of dispersal and spatial heterogeneity of the environment on the total population size of a single species. Such a consideration is not just out of curiosity, but rather aims at interesting connections with the issue of invasions of species. The connection will become clear in later sections. More precisely, we consider

$$
\begin{cases}
\mu\Delta\theta + \theta\big[m(x) - \theta\big] = 0 & \text{in } \Omega, \qquad \theta > 0 \text{ in } \Omega, \\
\dfrac{\partial\theta}{\partial n} = 0 & \text{on } \partial\Omega,
\end{cases}
\tag{5.9}
$$

where the migration rate μ is assumed to be a positive constant and the function $\theta = \theta(x, \mu)$ represents the density of the species at location x.

Though most of our analysis covers the case $\int_\Omega m(x)\,dx \leq 0$, throughout this subsection, for the sake of clarity, we posit that

(**A3**) $m(x)$ is nonconstant, bounded and measurable, and $\int_\Omega m(x)\,dx > 0$.

For solutions of (5.9), most of the following results are well known.

Theorem 5.3.1 *Suppose that assumption* (A3) *holds.*

(a) *For every $\mu > 0$, the problem* (5.9) *has a unique positive solution $\theta(x, \mu)$ such that $\theta \in W^{2,p}(\Omega)$ for every $p \geq 1$.*
(b) *As $\mu \to 0+$, $\theta(x, \mu) \to m_+(x)$ in $L^p(\Omega)$ for every $p \geq 1$, where $m_+(x) = \sup\{m(x), 0\}$; as $\mu \to \infty$, $\theta(x, \mu) \to \frac{1}{|\Omega|}\int_\Omega m(x)\,dx$ in $W^{2,p}(\Omega)$ for every $p \geq 1$.*
(c) *If $m(x)$ is Hölder continuous in $\overline{\Omega}$, then $\theta \in C^2(\overline{\Omega})$. Moreover, $\theta(x, \mu) \to m_+(x)$ in $L^\infty(\Omega)$ as $\mu \to 0$, and $\theta(x, \mu) \to \frac{1}{|\Omega|}\int_\Omega m(x)\,dx$ in $C^2(\overline{\Omega})$ as $\mu \to \infty$.*

We refer the proofs of parts (a) and (c) to [82] and the references therein. In view of part (b) of Theorem 5.3.1, it is natural to introduce the function

$$F(\mu) \equiv \begin{cases} \displaystyle\int_{\Omega} m_+(x)\,\mathrm{d}x, & \mu = 0, \\[2ex] \displaystyle\int_{\Omega} \theta(x,\mu)\,\mathrm{d}x, & \mu > 0, \\[2ex] \displaystyle\int_{\Omega} m(x)\,\mathrm{d}x, & \mu = \infty, \end{cases}$$

which can be interpreted as the total population size of the species. By assumption (A3) and part (b) of Theorem 5.3.1, F is a continuous, positive function in $[0, \infty]$.

If the spatial environment is homogeneous, i.e., $m(x)$ is equal to some positive constant $\overline{m}$, then $\theta(x,\mu) \equiv \overline{m}$ is the unique positive solution of (5.9) for every $\mu > 0$. In this case, the total population size of the species is given by $F(\mu) = |\Omega|\overline{m}$, which is independent of μ. However, if the spatial environment is heterogeneous, i.e., $m(x)$ is a nonconstant function, the story changes dramatically:

Theorem 5.3.2 ([82]) *Suppose that assumption* (A3) *holds.*

(a) *The function $F(\mu)$ satisfies $F(\mu) > F(\infty)$ for every $\mu \in (0, \infty)$.*
(b) *If $m(x) \geq 0$ in Ω, then for every $\mu \in (0, \infty)$, the function $F(\mu)$ satisfies*

$$F(0) = F(\infty) < F(\mu).$$

Part (a) of Theorem 5.3.2 implies that spatial heterogeneity increases the population size of species. To make this assertion precise, set $\overline{m} = \int_{\Omega} m(x)\,\mathrm{d}x/|\Omega|$, and write $F = F(\mu, m)$ instead of $F(\mu)$ to indicate the dependence of F on the function m. Part (a) implies that $F(\mu, m) > F(\mu, \overline{m})$ for every $\mu > 0$. In other words, given any $\mu > 0$ and any function g with $\int_{\Omega} g(x)\,\mathrm{d}x = 0$ and $g \not\equiv 0$, we have $F(\mu, \overline{m} + \lambda g) > F(\mu, \overline{m})$ for every $\lambda \neq 0$. Hence, with the dispersal rate being fixed, the population size $F(\mu, \overline{m} + \lambda g)$, as a function of λ, attains a strict absolute minimum at $\lambda = 0$.

Part (b) of Theorem 5.3.2 implies that when $m(x)$ is nonnegative, the total population size is minimized at $\mu = 0$ and $\mu = \infty$, and maximized at some intermediate value μ^*. The value of μ^* is determined by the habitat Ω and $m(x)$.

It will be of interest to understand the precise shape of $F(\mu)$ due to its crucial role in invasion of species. One natural conjecture is that $F(\mu)$ has a unique local maximum (and thus it must be the global maximum) in $(-\infty, +\infty)$. However, this conjecture is false (V. Hutson, personal communication) even for the case when $m(x)$ is a perturbation of positive constants.

If the function $m(x)$ changes sign, $F(\mu) > F(\infty)$ still holds for every $\mu \in [0, \infty)$. Hence, F is minimized at $\mu = \infty$, but is no longer minimized at $\mu = 0$. Interestingly, when Ω is an interval, the maximum of F is still attained at some intermediate dispersal rate:

Theorem 5.3.3 ([82]) *Suppose that $\Omega = (0, 1)$, $m \in C^2[0, 1]$, m changes sign, and $m(x) = 0$ has only nondegenerate roots in $[0, 1]$. Then there exist* *positive constants μ_0 and c_0 such that $F(\mu) - F(0) \geq c_0 \mu^{\frac{2}{3}}$ for every $\mu \in$ $(0, \mu_0)$. Thus,*

$$F(\infty) < F(0) < \sup_{0 < \mu < \infty} F(\mu).$$

Theorems 5.3.2 and 5.3.3 suggest that the total population size of species is usually maximized at some intermediate migration rate, and this fact has interesting applications to multiple species in the context of ecological invasions [82].

Another interesting problem is to maximize the total population for certain classes of $m(x)$ (with μ being fixed). Recent investigations (V. Hutson, personal communication) show that the results can be different for one-dimensional and higher dimensional domains. For example, given any $m_0 > 0$, the total population size, as a function of m, is a bounded (nonlinear) functional in the set $\{m \in L^\infty(\Omega) : m \geq 0 \text{ in } \Omega, \int_\Omega m = m_0|\Omega|\}$ for one-dimensional habitats, but is unbounded for high dimensional habitats. This suggests that one needs to consider the problem of maximizing $\int_\Omega \theta$ in a smaller set such as $\{m : 0 \leq m \leq 1 \text{ in } \Omega, \int_\Omega m = m_0|\Omega|\}$ with $m_0 \in (0, 1)$. The restriction that m, which is the per capita rate of increase when the species is rare, is uniformly bounded seems biologically reasonable.

From the control point of view, if we regard $\int_\Omega \theta$ as gain and $\|m\|_{L^2(\Omega)}$ as cost, it is also interesting to maximize the "profit" such as $\int_\Omega \theta - \|m\|_{L^2(\Omega)}$ instead of the total population size alone.

5.3.2 Biased Movement of Species

Throughout this subsection, we assume that $\lambda > \lambda_*(\alpha)$ so that the elliptic problem

$$\begin{cases} \nabla \cdot [\nabla \tilde{u} - \alpha \tilde{u} \nabla m] + \lambda(m - \tilde{u})\tilde{u} = 0 & \text{in } \Omega, \\ \dfrac{\partial \tilde{u}}{\partial n} - \alpha \tilde{u} \dfrac{\partial m}{\partial n} = 0 & \text{on } \partial\Omega \end{cases} \tag{5.10}$$

has a unique positive solution, which we denote as $\tilde{u} = \tilde{u}(x, \alpha)$. We will study various properties of $\tilde{u}$, including the total population size, the maximum value of $\tilde{u}$, and the concentration of $\tilde{u}$ at the maximum of $m(x)$.

When α is positive and small, the movement along the gradient of resources is relatively weak in comparison to random dispersal. For every fixed λ, it is shown [31] that increasing α is always beneficial to the persistence of the species for any convex habitat, but can be harmful in certain nonconvex habitats. This implies that the direction of selection of conditional dispersal can depend on the geometry of the habitat.

What happens if α increases and becomes suitably large? It turns out that large α always ensure the persistence of the species, independent of the geometry of the habitat. In other words, a conditional dispersal strategy can be advantageous over unconditional ones since it uses more information about resources. However, over pursuing this extra information can also be costly. Recently, we found that if the set of critical points of $m(x)$ has Lebesgue measure zero, then the total population size tends to zero as $\alpha \to \infty$ (see [22]), i.e.,

$$\lim_{\alpha \to \infty} \int_\Omega \tilde{u}(x; \alpha)\, \mathrm{d}x = 0. \tag{5.11}$$

In other words, the species has to pay a price if the species is "smart" but gets too "greedy," as small populations are more likely to go extinct. However, the population density may not approach zero everywhere:

Theorem 5.3.4 *Suppose that m is nonconstant.*

(a) *If $\alpha \leq 1/\max_{\bar{\Omega}} m$, then*

$$\tilde{u}(x) < \max_\Omega m\, e^{\alpha[m(x) - \max_{\bar{\Omega}} m]}$$

for every $x \in \bar{\Omega}$. In particular, $\max_{\bar{\Omega}} \tilde{u} < \max_{\bar{\Omega}} m$.
(b) *If $m(x) > 0$ in $\bar{\Omega}$ and $\alpha \geq 1/\min_{\bar{\Omega}} m$, then*

$$\tilde{u}(x) > \max_\Omega m\, e^{\alpha[m(x) - \max_{\bar{\Omega}} m]} \tag{5.12}$$

for every $x \in \bar{\Omega}$. In particular, $\max_{\bar{\Omega}} \tilde{u} > \max_{\bar{\Omega}} m$.

Hence, for sufficiently large α, if $m(x) > 0$ in $\bar{\Omega}$, then solutions of (5.10) are concentrated at the global maxima of $m(x)$. Another interesting consequence of Theorem 5.3.4 is that $\max_{\bar{\Omega}} \tilde{u} - \max_{\bar{\Omega}} m$, as a function of α, changes sign as α varies from zero to ∞. This fact will have interesting applications to the invasion of rare species in the study of the evolution of conditional dispersal. We conjecture that $\max_{\bar{\Omega}} \tilde{u} - \max_{\bar{\Omega}} m$ changes sign exactly once for all $\alpha \in (0, +\infty)$.

The case of $m(x)$ being positive everywhere corresponds to the situation when there are only sources all over the whole habitat, and it is natural to enquire what happens when environments comprise both sources and sinks:

Conjecture. For any nonconstant function $m(x)$, the solution of (5.10) is always concentrated at every local positive maximum of $m(x)$.

This conjecture is true when Ω is an interval and $m(x)$ has no critical points. More precisely, it is shown [22] that if $m'(x) > 0$ in $[0, 1]$, then for sufficiently large α, $\tilde{u}'(x) > 0$ in $[0, 1]$, $\tilde{u}(x) \to 0$ uniformly in $[0, c]$ for every $c \in (0, 1)$ as $\alpha \to \infty$, and

$$u(1) \geq \int_0^1 m > 0$$

for large α, provided that $\int_0^1 m > 0$. If Ω is an interval and $m(x)$ is not monotone, under suitable conditions we can show that concentration of $\tilde{u}$ is only possible at critical points of $m(x)$. In this connection, we make the following assumption.

(A4) Suppose that $\Omega = (0, 1)$, $m_x(0) \geq 0 \geq m_x(1)$, and $m(x)$ has finitely many critical points in $[0, 1]$, denoted by $\{x_1, ..., x_k\}$.

Theorem 5.3.5 ([22]) *Suppose that* (A4) *holds and* $\Omega = (0,1)$. *Then* $\tilde{u}(x) \to$ 0 *uniformly in every compact subset of* $[0,1] \setminus \{x_1, ..., x_k\}$ *as* $\alpha \to \infty$. *In particular,* $\tilde{u}(x) \to 0$ *pointwise for every* $x \in [0,1] \setminus \{x_1, \ldots, x_k\}$ *as* $\alpha \to \infty$.

5.4 Evolution of Dispersal and Related Topics

It is natural to inquire which patterns of dispersal can confer some selective or ecological advantage, see [4, 31, 35, 61, 68, 94] and references therein. Unconditional dispersal does not depend on habitat quality or population density, while conditional dispersal does depend on some or all of such factors. For instance, passive diffusion as considered by Dockery et al. [35] and Hutson et al. [68] is a type of unconditional dispersal. Diffusion combined with directed movement along resource gradients, as considered by Belgacem and Cosner [4] and Cosner and Lou [31], is an example of conditional dispersal, because the bias in the direction of dispersal depends on the spatial distribution of resources. It was shown in both patch and diffusion models (McPeek and Holt [94]; Dockery et al. [35]) that for unconditional dispersal in spatially varying but temporally constant environments slower dispersal rates is selected, and the results in Hastings [54] essentially say that decreasing dispersal rate increases some measures of fitness in logistic models. However, for unconditional dispersal in spatially and temporally varying environments faster dispersal rates may be selected in both patch models [94] and diffusion models [68]. In this section we focus on the effects of unconditional dispersal on the dynamics of two competing species in a spatially varying (but temporally constant) environment.

The semilinear parabolic system

$$\frac{\partial u}{\partial t} = \mu \Delta u + u[m(x) - u - bv] \quad \text{in } \Omega \times (0, \infty),$$

$$\frac{\partial v}{\partial t} = \nu \Delta v + v[m(x) - cu - v] \quad \text{in } \Omega \times (0, \infty), \quad (5.13)$$

$$\frac{\partial u}{\partial n} = \frac{\partial v}{\partial n} = 0 \quad \text{on } \partial\Omega \times (0, \infty)$$

models two species that are competing for the same resources, where $u(x,t)$ and $v(x,t)$ represent the population densities of competing species 1 and 2 with respective dispersal rates μ and ν, the function $m(x)$ represents their common intrinsic growth rate, and b and c are inter-specific competition coefficients. We shall assume that μ, ν, b, and c are positive constants, and $u(x,0)$ and $v(x,0)$ are non-negative functions that are not identically equal to zero.

If we assume that the initial data $u(x,0)$ and $v(x,0)$ are non-negative and not identically zero, then by maximum principle [45, 110], $u(x,t) > 0$ and $v(x,t) > 0$ for every $x \in \overline{\Omega}$ and every $t > 0$. Moreover, $u(x,t)$ and $v(x,t)$ are classical solutions of (5.13) and exist for all time $t > 0$. Of particular interest are the dynamics and coexistence states of (5.13). We say that a steady state (u_e, v_e) of (5.13) is a *coexistence state* if both components are positive, and it is a *semi-trivial state* if one component is positive and the other is zero. Under assumption (A3), (5.13) has two semi-trivial states, denoted by $(\theta(\cdot, \mu), 0)$ and $(0, \theta(\cdot, \nu))$ for every $\mu > 0$ and every $\nu > 0$, where $\theta(\cdot, \mu)$ is the unique positive solution of (5.9).

For the last two decades there has been tremendous interest, by both mathematicians and ecologists, in two-species Lotka-Volterra competition models with spatially homogeneous or heterogeneous interactions, see [9, 10, 11, 12, 18, 19, 28, 29, 33, 35, 38, 39, 40, 41, 46, 55, 59, 62, 63, 64, 65, 66, 71, 75, 80, 81, 93, 96, 107, 109] and references therein. For competition models with density-dependent diffusion, we refer to [27, 69, 83, 87, 88, 89, 93, 95, 97, 98, 116] and references therein.

5.4.1 The Slower Diffuser Wins

Consider the case when $b = c = 1$. It is shown in [35] that if $\mu < \nu$, then $(\theta(\cdot, \mu), 0)$ is globally asymptotically stable among all nonnegative nontrivial initial data. In other words, *the slower diffuser wins*. By symmetry, a similar conclusion holds when $\mu > \nu$. In particular, (5.13) has no coexistence states if $\mu \neq \nu$.

Why is the slower diffuser always the winner? Such a phenomenon is a little surprising at the first look: if $\mu = \nu = 0$, it is clear that neither species will die out; in fact, the two species will coexist since they are identical. However, as soon as the diffusion is turned on for both species, the slower diffuser becomes the eventual winner as time evolves. This shows that the PDE dynamics are dramatically different from the ODE dynamics.

A possible biological explanation is that as time evolves, the effective growth rate $a(x, t) := m(x) - u(x, t) - v(x, t)$ for both species will eventually change sign in the habitat Ω. The slower diffuser keeps relatively low density in the region where $a(x, t)$ is negative, which seems to help it gain some competitive advantage.

A challenging open problem is whether the slowest diffuser still wins the competition in the context of N competing species with $N \geq 3$. One major new mathematical difficulty in solving this open problem is that competition models for three or more species are not monotone systems. For recent progress on patch models, see [74] and references therein.

Similar results hold true for the case of nonlocal dispersions, and we refer to [67] for the details. However, when the intrinsic growth rate varies periodically in time, it is shown in [94] for patch models and in [68] for diffusion models that the slower diffuser may not always be the winner. That is to say, the effects of spatial and temporal variations can be quite different. Clearly, understanding the effects of temporal variation is mathematically very challenging and biologically rich.

5.4.2 ODE Dynamics vs. PDE Dynamics

To motivate our discussion, we assume for the moment that $0 < b, c < 1$. If $m(x) \equiv \overline{m}$ for some positive constant $\overline{m}$, then every solution (u, v) of (5.13) converges to $\left(\frac{1-b}{1-bc}\overline{m}, \frac{1-c}{1-bc}\overline{m} \right)$ for all diffusion rates μ, ν, and arbitrary initial data. In other words, the PDE dynamics are no different from the corresponding ODE ones. In fact, for any autonomous reaction–diffusion system with no-flux boundary conditions, if the initial data are constants, then the corresponding PDE solutions remain spatially homogeneous. In other words, the dynamics of ODEs are "embedded" in the dynamics of corresponding PDEs (this is different from Dirichlet or Robin conditions that induces a hidden spatially inhomogeneous effect). However, for nonautonomous reaction–diffusion systems with no-flux boundary conditions, things can go quite differently. To this end, we investigate the asymptotic behavior of solutions of (5.13) when $m(x)$ is a nonconstant positive function and $0 < b, c < 1$.

If $\mu = \nu = 0$ in (5.13), we can regard (5.13) as a system of two ordinary differential equations, solutions of which converge to $\left(\frac{1-b}{1-bc}m(x), \frac{1-c}{1-bc}m(x) \right)$ for every $x \in \Omega$. In other words, the kinetic system of (5.13) has a unique, globally asymptotically stable positive equilibrium (depending on x). It is shown in [65] that if μ and ν are sufficiently small, the function $m(x)$ is positive, and $0 < b, c < 1$, then (5.13) has a unique, globally asymptotic stable positive steady state (u^*, v^*). Moreover, (u^*, v^*) converges to $\left(\frac{1-b}{1-bc}m(x), \frac{1-c}{1-bc}m(x) \right)$ in $L^\infty(\Omega)$ as $\mu \to 0$ and $\nu \to 0$. That is, when both μ and ν are small, the dynamics of (5.13) behave similarly to those of (5.13) with $\mu = \nu = 0$. It would be very interesting to discover whether this type of result holds for other important classes of interacting species, for example, for $N(\geq 3)$ competing species.

When both μ and ν are sufficiently large, it is not difficult to see that (5.13) again has a unique, globally asymptotically stable positive steady state (u^{**}, v^{**}). Moreover, (u^{**}, v^{**}) converges to $\left(\frac{1-b}{1-bc} \frac{1}{|\Omega|} \int_\Omega m, \frac{1-c}{1-bc} \frac{1}{|\Omega|} \int_\Omega m \right)$ as $\mu \to \infty$ and $\nu \to \infty$. In fact, when μ and ν are sufficiently large, the dynamics of (5.13) can be approximated by that of the corresponding "spatially averaged" ODE. Consult [23, 30, 49, 51] and references therein for the connection between the dynamics of PDEs with large diffusion and that of corresponding "spatially averaged" ODEs.

Hence, it seems reasonable to expect that for other ranges of dispersal rates, the dynamics of (5.13) should still be well behaved; *e.g.*, the two competing species can coexist, as in the case when μ and ν are both small or both large, or in the case when $m(x)$ is equal to some positive constant. However, as we shall see later, given any nonconstant positive function $m(x)$, there exists a set of parameters $b, c \in (0, 1)$ and $\mu, \nu > 0$ such that one of the semi-trivial steady states of (5.13) is the global attractor of (5.13). Therefore, the joint action of spatial heterogeneity and diffusion can drive one of the species to extinction. Such "diffusion-driven extinction" or "diffusion-driven blowup" phenomena have been studied in [11, 70, 86, 100, 105, 107, 122]. It was proved in [70, 105] that there exist some initial data such that without diffusion species 1 drives species 2 to extinction, whereas the opposite holds for certain dispersal rates. In [107] by numerical investigations Pacala and Roughgarden made the following counterintuitive observations: (1) without diffusion one competitor can not invade the other at any location but with diffusion invasion is possible; (2) without diffusion one competitor can invade the other at every location but with diffusion invasion may fail. These observations were explained and rigorously verified later on by Cantrell and Cosner in [11]. In [82], the spatial heterogeneity is also incorporated in the nonlinearity and the extinction result holds for arbitrary initial conditions.

We start by studying the joint effects of diffusion and spatial heterogeneity on the invasion of the species 2 when it is rare. The case of species 1 is similar. Mathematically, this is equivalent to studying the stability of the semi-trivial steady state $(\theta(x, \mu), 0)$ of (5.13).

For the rest of this subsection, we focus on the case $0 < c < 1$. The stability of $(\theta, 0)$ when $c \in (0, 1)$ can be described by

Theorem 5.4.1 *If assumption (A3) holds and $m(x)$ is nonnegative, then there exists some constant $c_* = c_*(m, \Omega) \in (0, 1)$ such that the following results hold.*

(a) *For every $c \in (0, c_*)$, the steady state $(\theta, 0)$ is unstable when $\mu > 0$ and $\nu > 0$.*

(b) *For every $c \in (c_*, 1)$, there exists $\bar{\nu} = \bar{\nu}(c, m, \Omega) > 0$ such that (1) for every $\nu \in (0, \bar{\nu})$, the steady state $(\theta, 0)$ is unstable when $\mu > 0$; (2) for every $\nu > \bar{\nu}$, the steady state $(\theta, 0)$ changes stability at least twice as μ increases from 0 to ν.*

A much more detailed description of the region in the $\mu\nu$-plane where $(\theta, 0)$ is stable is given in [82], Sect. 4. For the case when $m(x)$ changes sign, the stability of $(\theta, 0)$ is somewhat different. We refer to [82] for details.

The most interesting region is where $c_* < c < 1$ and $\nu > \bar{\nu}$, where we have the following:

(a) If $b > 1$, it is well known that without dispersal, species 2 always drives species 1 to extinction. However, with dispersal, for some ranges of dispersal rates, species 2 may fail to invade when rare.

(b) If $b < 1$, it is well known that, without dispersal, species 1 always coexists with species 2. Surprisingly, for certain dispersal rates, species 1 is able to drive species 2 to extinction for arbitrary initial conditions.

For every $c > 0$, define

$$\Sigma_c = \{(\mu, \nu) \in (0, \infty) \times (0, \infty) : (\theta, 0) \text{ is linearly stable}\}. \tag{5.14}$$

Theorem 5.4.2 *If assumption (A3) holds and $m(x)$ is nonnegative, then for every $c \in (c_*, 1)$, there exists $b_* = b_*(c, \Omega, m) \in (0, 1]$ such that if $b \in (0, b_*]$ and $(\mu, \nu) \in \Sigma_c$, then $(\theta, 0)$ is globally asymptotically stable.*

We conjecture that Theorem 5.4.2 holds with $b_* = 1$:

Conjecture. If assumption (A3) holds, $m(x)$ is nonnegative, and $b = 1$, then for every $c \in (c_*, 1)$ and $(\mu, \nu) \in \Sigma_c$, $(\theta, 0)$ is globally asymptotically stable.

By Theorem 5.4.1, the set Σ_c is nonempty for every $c \in (c_*, 1)$. It is not difficult to see that $\Sigma_{c_1} \subset \Sigma_{c_2}$ for any $c_1 < c_2$ with $c_1, c_2 \in (c_*, 1)$. In fact, the set Σ_c converges to the set $\{(\mu, \nu) : 0 < \mu < \nu\}$ as $c \to 1-$, and this conjecture gives another perspective on why the slower diffuser wins for the case when $b = c = 1$.

5.5 Similar Competing Species: Invasion and Coexistence

To motivate our discussions, we start with the semilinear parabolic system

$$
\begin{aligned}
u_t &= \mu \Delta u + u \left[m(x) - u - v \right] && \text{in } \Omega \times (0, \infty), \\
v_t &= \mu \Delta v + v \left[m(x) - u - v \right] && \text{in } \Omega \times (0, \infty), \\
\frac{\partial u}{\partial n} &= \frac{\partial v}{\partial n} = 0 && \text{on } \partial \Omega \times (0, \infty).
\end{aligned}
\tag{5.15}
$$

Clearly, the two species are identical in all aspects except their initial conditions. By assumption (A3), (5.15) has a family of coexistence states, given by $\{(s\theta, (1-s)\theta) : 0 < s < 1\}$, which attracts all solutions with nonnegative nontrivial initial data, where $\theta = \theta(\cdot, \mu)$ is the unique positive solution of (5.9). Moreover, for any nonnegative nontrivial initial data, the solution of (5.15) converges to $(s_0 \theta, (1 - s_0)\theta)$ for some $s_0 \in (0, 1)$, where s_0 depends on the initial data. It is an interesting problem to determine how this structure changes under small perturbations:

Question. What happens when the two species are slightly different, that is, when system (5.15) is perturbed?

Biologically, suppose that random mutation produces a different phenotype of species, which is slightly different from the original species, e.g., different colors for butterflies, different sizes of wings for birds, etc. It is fairly reasonable to expect that these two species will have to compete for rather similar (if not exactly the same) resources. The major concern is whether the mutant can invade when rare; if so, will the invading species force the extinction of the resident species or coexist with it?

Mathematically, the question leads to the study of various perturbations of system (5.15). Several interesting and surprising phenomena will be revealed in the next few subsections using this approach. What is particularly attractive is that this perturbation method can yield lots of information on competing species for *intermediate* values of diffusion rates, which is in general hard to find.

5.5.1 Effects of Resources

In [66] a perturbation of the intrinsic growth rate in (5.15) is considered. The system studied there has the form

$$u_t = \mu \Delta u + u[m(x) + \tau g(x) - u - v] \qquad \text{in } \Omega \times (0, \infty), \qquad (5.16a)$$

$$v_t = \mu \Delta v + v[m(x) - u - v] \qquad \text{in } \Omega \times (0, \infty), \qquad (5.16b)$$

$$\frac{\partial u}{\partial n} = \frac{\partial v}{\partial n} = 0 \qquad \text{on } \Omega \times (0, \infty). \qquad (5.16c)$$

The two species are almost identical except for their intrinsic growth rates, which differ by a function of $\tau g(x)$, where τ is a positive constant and $g(x)$ is a smooth function. In this situation, some new phenomena are discovered. By (A3), if $\int_\Omega g > 0$, (5.16) has two semi-trivial states in the form of $(\tilde{u}, 0)$ and $(0, \theta)$ for every $\mu > 0$.

It was shown in [63] that if $\int_\Omega g > 0$, then for large enough μ, $(\tilde{u}, 0)$ is a global attractor. In particular, $(\tilde{u}, 0)$ is asymptotically stable and $(0, \theta(x, \mu))$ is unstable. That is, the species u always drives the species v to extinction, no matter what the initial data may be. On the other hand, [63] also demonstrates that if $m_+ - (m + \tau g)_+$ (a_+ denotes the positive part of function $a(x)$) changes sign, then for small enough μ both semi-trivial states $(\tilde{u}, 0)$ and $(0, \theta)$ of (5.16) are unstable. This in turn implies that there is at least one stable coexistence state of (5.16) for small μ.

The simplest interpretation of these results suggest that as μ decreases from a large value, a branch of coexistence states of (5.16) bifurcates from $(\tilde{u}, 0)$ at some value μ_0 and remains in the interior of the positive cone for all $\mu < \mu_0$. Surprisingly, numerical computations show that this is not always the case. For reasonable choices of $m(x)$ and $g(x)$, the branch of coexistence states bifurcating from $(\tilde{u}, 0)$ at μ_0 will connect to $(0, \theta)$ at μ_1 for some $\mu_1 < \mu_0$, and $(0, \theta)$ becomes globally attracting for some range of μ as μ decreases. Eventually, another branch of coexistence states of (5.16) bifurcates from $(0, \theta)$ at μ_2 for some $\mu_2 < \mu_1$, and remains in the positive cone for the rest of μ. This is surprising for the situation $\int_\Omega g > 0$ because species u was chosen to have better average reproductive rate than species v.

The study in [66] on the stability of $(0, \theta)$ partially confirms the numerical results, i.e., the stability can change and indeed can do so more than once as the diffusion rate μ is varied:

Theorem 5.5.1 *Suppose that* (A3) *holds and* $\int_\Omega g(x) > 0 > \int_\Omega g(x) m_+^2(x)$. *Then there exists a unique* $\tau_0 > 0$ *such that*

(i) $\tau > \tau_0$ *implies that* $(0, \theta)$ *is unstable for any* $\mu > 0$;
(ii) $\tau < \tau_0$ *implies that* $(0, \theta)$ *changes stability at least once as* μ *varies from zero to infinity. It changes stability at least twice, provided* g *and* m *are both positive on a nonempty subset of* Ω.

The stability of $(0, \theta)$ means that the first species with low density u can not invade. Theorem 5.5.1 qualitatively illustrates how the invasion of rare species relies on both dispersal rate and the difference between its intrinsic growth rate and that of resident species. We conjecture that for case (ii), whenever $(0, \theta)$ is locally asymptotically stable, it is globally asymptotically stable.

For small τ, it is further shown in [66] that for any fixed positive integer k, one can choose the function g from an open set of possibilities such that $(\tilde{u}, 0)$ and $(0, \theta)$ exchange their stability at least k times as the diffusion rate μ varies over $(0, \infty)$. As a consequence, there are at least k branches of coexistence states of (5.16), which connect $(\tilde{u}, 0)$ and $(0, \theta)$. Moreover, coexistent states are unique and globally asymptotically stable if they exist. This confirms the previous conjecture, at least for the case when τ is positive and small. Biologically, this implies that with small variations of the phenotype, the stability of the two species varies with diffusion rate in a very complex manner, and it is unpredictable which species will survive. It also suggests that mutation leads to multiple opportunities for coexistence and thus potentially for speciation.

For small τ, which is more biologically realistic, these switches of stabilities between the two semi-trivial steady states happen within narrow regions of μ. But as τ increases, such regions widen up. Nonetheless, it suggests that there is no surprise in finding a large range of coexisting phenotypes that differ only in one small manner, e.g., the manner in which they utilize the resources of the environment.

When $m(x)$ is a positive constant, it was recently shown in [48] that for any $\tau > 0$ and any nontrivial function g satisfying $\int_\Omega g(x)\,\mathrm{d}x = 0$, for sufficiently large μ, $v(x, t) \to 0$ as $t \to \infty$. In other words, when both competing species have sufficiently large dispersal rates, the species with larger spatial variation in its intrinsic growth rate has the better chance of survival. We also refer to [48] for some related work on patch models, and to [85] for a similar approach in the study of evolution of gene frequency under joint forces of migration and selection.

5.5.2 Effects of Interspecific Competition

We consider the system

$$u_t = \mu \Delta u + u\left\{m(x) - u - [1 + \tau g(x)]v\right\} \qquad \text{in } \Omega \times (0, \infty), \qquad (5.17a)$$

$$v_t = \mu \Delta v + v\left\{m(x) - v - [1 + \tau h(x)]u\right\} \qquad \text{in } \Omega \times (0, \infty), \qquad (5.17b)$$

$$\frac{\partial u}{\partial n} = \frac{\partial v}{\partial n} = 0 \qquad \text{on } \Omega \times (0, \infty), \qquad (5.17c)$$

where the two species are identical except for their interspecific competition rates, which are given by $1 + \tau g(x)$ and $1 + \tau h(x)$, respectively. Here τ is a positive constant and $g(x)$, $h(x)$ are two smooth functions. In comparison to (5.16), some new structure of coexistence equilibria of (5.17) is found.

Let u, v denote the densities of the resident species and the mutant, respectively. Define

$$\Omega_+ = \{x \in \Omega : g(x) > 0 > h(x)\}, \qquad \Omega_- = \{x \in \Omega : g(x) < 0 < h(x)\}.$$

In Ω_+, the mutant has the competitive advantage. If diffusion is not present, then the mutant not only can invade, but also goes to fixation, i.e., it forces the extinction of the original phenotype. The outcome is reversed in Ω_-. Biologically, when diffusion is present, it would be interesting to determine whether the mutant can coexist with the original species, and/or whether the mutant can invade. Clearly, such phenomena can occur only when spatial heterogeneity is involved, since the answer is negative if both g and h are constant functions.

Since g and h can be rather general, the dynamics of (5.17) and the structures of coexistence states can potentially be very complicated, and it seems impossible to find any simple criteria that could characterize them. However, quite amazingly, for small τ, the dynamics and coexistence states of (5.17) essentially depend on two scalar functions of $\mu \in (0, \infty)$ defined by

$$G(\mu) = \int_\Omega g(x)\theta^3(x, \mu)\, dx, \tag{5.18a}$$

$$H(\mu) = \int_\Omega h(x)\theta^3(x, \mu)\, dx, \tag{5.18b}$$

where $\theta(x, \mu)$ is the unique positive solution of (5.9). The following theorem shows how G and H determine the structure of coexistence states and their stability.

Theorem 5.5.2 ([84]) *Assume that functions G and H have no common roots. Let μ_1 and μ_2 be two consecutive roots of the function GH and assume that they are both simple roots.*

(i) *If $GH < 0$ in (μ_1, μ_2), then for $\mu \in [\mu_1, \mu_2]$, system (5.17) has no coexistence states provided that τ is small and positive.*

(ii) *If $GH > 0$ in (μ_1, μ_2), then for each sufficiently small $\tau > 0$ there exist numbers $\underline{\mu} = \underline{\mu}(\tau) \approx \mu_1$, $\overline{\mu} = \overline{\mu}(\tau) \approx \mu_2$ such that for each $\mu \in (\underline{\mu}, \overline{\mu})$, (5.17) has a unique coexistence state, and $(u(\underline{\mu}), v(\underline{\mu}))$, $(u(\overline{\mu}), v(\overline{\mu}))$ are semi-trivial states of (5.17). Moreover, the coexistence state is stable if and only if both $G(\mu)$ and $H(\mu)$ are negative in (μ_1, μ_2).*

In statement (ii), the following two scenarios can occur:

(a) $(u(\underline{\mu}), v(\underline{\mu}))$ and $(u(\overline{\mu}), v(\overline{\mu}))$ are semi-trivial states of the same type, that is, each of them equals $(\theta, 0)$ (at the corresponding value of μ) or each of them equals $(0, \theta)$,

(b) $(u(\underline{\mu}), v(\underline{\mu}))$ and $(u(\overline{\mu}), v(\overline{\mu}))$ are of different types: one of them equals $(\theta, 0)$ and the other one equals $(0, \theta)$.

In the case (a) we call the curve $\{(u(\mu), v(\mu)) : \mu \in (\underline{\mu}, \overline{\mu})\}$ a *branch* between $\underline{\mu}$ and $\overline{\mu}$; in the case (b) we call it a *loop*. If the coexistence states on the branch or loop are stable we call it a *stable branch* or a *stable loop*, respectively. It turns out that (5.17) can have an arbitrarily high number of stable loops and branches.

Theorem 5.5.3 ([84]) *Suppose that* (A3) *holds,* $m \in C^\gamma(\overline{\Omega})$ *and* $m_+^3 \notin C^{\gamma+1}(\overline{\Omega})$ *for some* $\gamma > 0$. *Then, for any given positive integers* l *and* b, *there exist smooth functions* g *and* h *such that* (5.17) *has at least* l *stable loops and at least* b *stable branches for each sufficiently small* $\tau > 0$.

Theorem 5.5.3 reveals complex and intriguing effects of diffusion and spatial heterogeneity of the environment on the invasion of rare species and coexistence of interacting species. The existence of (stable) loops appears to be a new phenomenon, as it does not occur in the model studied in [66]. Also, in contrast to the results of [66], the range of coexistence in terms of μ, that is, the projection of a branch or loop of coexistence states onto the μ−axis, is of order $O(1)$ as $\tau \to 0$.

5.5.3 Effects of boundary condition

Recently, there has been considerable interest in how habitat edges change species interactions inside the habitat. For instance, in [16] Cantrell, Cosner and Fagan demonstrated how degrading the quality of the "matrix" habitat surrounding a habitat patch could reverse the nature of competitive two-species dynamics inside the patch so that a nominally "inferior" species out competes a "superior" one. However, the result in Theorem 6.2 of [16] does not rule out the possibility that the competitive advantage might switch back and forth between species 1 and species 2 a number of times before ultimately belonging to species 2. This observation raises a very interesting question, both ecologically and mathematically. Namely,

Question. Can a competitive advantage in such models reverse more than once as the level of degradation in the "matrix" habitat surrounding Ω increases?

To address such a question we examine the system

$$
\begin{aligned}
u_t &= \Delta u + u[1 + \epsilon g(x) - u - v] && \text{in} \quad \Omega \times (0, \infty), \\
v_t &= \Delta v + v[1 - u - v] && \text{in} \quad \Omega \times (0, \infty), \\
(1-s)\frac{\partial u}{\partial n} + su &= 0 = (1-s)\frac{\partial v}{\partial n} + sv && \text{on} \quad \partial\Omega \times (0, \infty),
\end{aligned}
\tag{5.19}
$$

where s ranges over $[0, 1]$. The perturbed system (5.19) is completely determined by the choice of the function g. It is shown in [20] that there exist functions g for which (5.19) exhibits multiple reversals of competitive advantage as the hostility of the "matrix" habitat surrounding Ω (which is measured by s) increases. This is distinct from the results in [16], since here the competitive advantage that one species has over the other is necessarily independent of initial configurations of species densities. In other words, when one of the species has the advantage, it competitively excludes the other over time. Moreover, under appropriate conditions on g, the regions in (ε, s) space in which one species excludes the other in (5.19) are bordered by values for which (5.19) admits a unique globally attracting componentwise positive equilibrium.

For other work on understanding the effects of habitat edges on species interactions, we refer to [15, 17, 43] and references therein. We also refer to [13] for some very interesting recent work concerning the effects of nonlinear boundary condition of the type

$$[1 - s(u)]\,\frac{\partial u}{\partial n} + s(u)u = 0$$

on the persistence of species in logistic population models, where $s(u) \in [0, 1]$ is a smooth function. It models the mechanism of inducing an Allee effect in the Glanville fritillary butterfly. Among other things, it is shown in [13] that under certain situations a branch of positive solutions satisfying $0 < u < 1$ connects with two trivial solutions $u \equiv 0$ and $u \equiv 1$ for *intermediate* values of dispersal rates. It will be of interest to see the effects of such nonlinear boundary conditions on multiple interacting species.

5.6 Evolution of Conditional Dispersal

A common underlying assumption of dispersal models is that dispersal is unconditional, i.e., organisms move at constant rates and in a nondirectional random manner. However, it is more probable that organisms can sense and respond effectively to local environmental cues by moving in the direction of increasingly favorable habitats. One of the simplest realistic approaches is to assume that organisms display taxis and can move up along the gradient of a local population growth rate. Diffusion combined with directed movement upward along resource gradients, as considered by Belgacem and Cosner [4] and Cosner and Lou [31], is an example of conditional dispersal, because the bias in the direction of dispersal depends on the spatial distribution of resources. In context of two-patch models, McPeek and Holt [94] showed in spatially varying but temporally constant environment conditional dispersal can be advantageous. In diffusion–advection models for a single population in a spatially varying but temporally constant environment, Belgacem and Cosner [4] and Cosner and Lou [31] showed that conditional dispersal involving both diffusion and directed movement up resource gradients can sometimes (but not always) make persistence more likely. For related work on balanced dispersal we refer to [37, 99] and references therein. We also refer to [14, 74] for recent progress on the evolution of conditional dispersal in patch models.

This section is devoted to studying conditional dispersal in context of competition between two populations that are ecologically identical except in their dispersal mechanisms.

We first assume that there is a random component to the dispersal of both competitors but that when resources are distributed in a spatially heterogeneous way, one of the competitors also has a tendency to move upward along the resource gradient while the other does not. Such directed motion introduces a drift or advection term into the diffusion equation. The dispersal of the two competitors may be described in terms of the fluxes $J_u = -\mu \nabla u + \alpha (\nabla m) u$ and $J_v = -\nu \nabla v$, where $\alpha \geq 0$. If we assume there is no flux across $\partial \Omega$, we obtain the competition model

$$u_t = \nabla \cdot [\mu \nabla u - \alpha u \nabla m] + (m - u - v)u$$
$$v_t = \nabla \cdot [\nu \nabla v] + (m - u - v)v \qquad (5.20)$$

in $\Omega \times (0, \infty)$, with boundary conditions

$$\mu \frac{\partial u}{\partial n} - \alpha u \frac{\partial m}{\partial n} = \frac{\partial v}{\partial n} = 0. \qquad (5.21)$$

When assumption (A3) holds, (5.20–5.21) has two semi-trivial states, denoted by $(\tilde{u}, 0)$ and $(0, \theta(\cdot, \nu))$, for every $\mu > 0$, $\nu > 0$, and $\alpha \geq 0$ (see [31]), where $\tilde{u}$ is the unique positive solution of (5.10). Throughout this section, we assume that m is twicely continuously differentiable.

5.6.1 Weak Advection

For fixed μ, ν with $\mu \neq \nu$, the dynamics of (5.20–5.21) is similar to that of (5.13) for sufficiently small α. More precisely, there exists some small positive constant $\alpha_0 = \alpha_0(\mu, \nu, \Omega, m)$ such that if $\alpha \in (0, \alpha_0)$, $(\tilde{u}, 0)$ is the global attractor of (5.20–5.21) among all nonnegative and nontrivial initial data if $\mu < \nu$, and $(0, \theta(\cdot, \nu))$ is the global attractor if $\mu > \nu$.

The case $\mu = \nu$ is quite delicate. This is due to the fact that (5.13) with $\mu = \nu$ is a degenerate system: it has a family of coexistence states, each of which is neutrally stable, and as a whole is a global attractor. For sufficiently small positive α, (5.20–5.21) can be viewed as a perturbation of (5.13).

For $\mu > 0$, define

$$\alpha^*(\mu) = \frac{\int_\Omega \theta(x, \mu) \nabla \theta(x, \mu) \cdot \nabla m(x) \, dx}{\int_\Omega |\nabla \theta(x, \mu)|^2 \, dx}.$$

As will be seen later, this quantity plays a crucial role in studying dynamics of (5.20–5.21) for small positive α.

For any $\mu_0 > 0$, $\mu_1, \nu_1 \in R^1$, and $\alpha_1 > 0$, let

$$(\mu, \nu, \alpha) = (\mu_0 + \mu_1 s + o(s), \mu_0 + \nu_1 s + o(s), \alpha_1 s + o(s)), \qquad (5.22)$$

where s is positive and small.

Theorem 5.6.1 *Suppose that (A3) holds and Ω is convex. Then we have*

(a) *For every $\mu > 0$, $\alpha^*(\mu) > 0$.*
(b) *Let μ, ν, α be given as in (5.22). If $\alpha_1 > (\mu_1 - \nu_1)/\alpha^*(\mu_0)$, then for positive small s, $(\tilde{u}, 0)$ is globally asymptotically stable. In particular, if $(\mu, \nu) = (\mu_0, \mu_0)$, $(\tilde{u}, 0)$ is globally asymptotically stable for small positive α.*

Theorem 5.6.1 is established in [21, 22], and it has some interesting consequences: e.g., for the case $\mu_1 > \nu_1$, it implies that the competitor that moves toward more favorable environments may have a competitive advantage even if it diffuses more rapidly than the other competitor. This is in strong contrast with the case in which both competitors disperse only by random diffusion, where the slower diffuser always wins. It means that the advantage gained from the directed movement upward resource gradients can compensate the disadvantage created by faster diffusion.

The convexity of Ω is needed in the proof of Theorem 5.6.1 to ensure that $\alpha^*(\mu) > 0$ for all $\mu > 0$, which allows us to exclude the possibility of coexistence states for small $\alpha > 0$. The proof of $\alpha^*(\mu) > 0$ is given in [21], where we applied the fact $\partial(|\nabla\theta|^2)/\partial n \le 0$ on $\partial\Omega$, which holds true for convex domains only. We should point out that the convexity assumption on domain Ω seems to be necessary, as shown by the following result [21].

Theorem 5.6.2 *Given any $\mu_0 > 0$, there exist nonconvex domain Ω and smooth function $m(x)$ such that*

(a) $\alpha^*(\mu_0) < 0$, *and* $\alpha^*(\mu)$ *changes sign at least once in* $(0, \mu_0)$;
(b) *Let* μ, ν, α *be given by (5.22). If* $\alpha_1 > (\mu_1 - \nu_1)/\alpha^*(\mu_0)$, *then for positive small* s, $(0, \theta(\cdot, \nu))$ *is globally asymptotically stable. In particular, if* $(\mu, \nu) = (\mu_0, \mu_0)$, $(0, \theta(\cdot, \mu_0))$ *is globally asymptotically stable for small positive* α.

For the case $\mu_1 < \nu_1$, part (b) of Theorem 5.6.2 implies that for certain nonconvex habitats, a slower diffuser which also moves toward more favorable environments may not have the competitive advantage. This is in strong contrast with both the case of a convex habitat and the case of $\alpha = 0$.

5.6.2 Advection Mediated Coexistence

In this subsection we are concerned with the much more interesting and challenging case wherein α is large, and show how strong advection can induce stable coexistence of competing species. In particular, we shall investigate the stability of $(\tilde{u}, 0)$ and $(0, \theta(\cdot, \nu))$, and the existence and qualitative properties of coexistence states. The stability of $(\tilde{u}, 0)$ and properties of coexistence states rely crucially on qualitative properties of $\tilde{u}$ given in Sect. 5.3.2.

For sufficiently large α, we have the following result.

Theorem 5.6.3 *Suppose that $\int_\Omega m > 0$ and the set of critical points of $m(x)$ has Lebesgue measure zero. Then for sufficiently large α, both $(\tilde{u}, 0)$ and $(0, \theta(\cdot, \nu))$ are unstable. Moreover, system (5.20–5.21) has at least one stable coexistence state.*

In fact, it can be shown that every coexistence state (u_α, v_α) of (5.20–5.21) satisfies $u_\alpha \to 0$ in $L^2(\Omega)$ and $v_\alpha \to \theta(\cdot, \nu)$ in $W^{2,2}(\Omega)$ as $\alpha \to \infty$. If we further assume that (A4) holds, $u_\alpha \to 0$ pointwise for every $x \in [0, 1] \setminus \{x_1, ..., x_k\}$.

Since (5.20–5.21) is a strongly monotone system, as in other competition models, the existence and stability of coexistence states follow from the instability of the two semi-trivial states and the theory of continuous monotone systems [32, 57, 62, 92, 118]. Furthermore, (5.20–5.21) has at least one asymptotically stable coexistence state [58]. For the discrete-time counterparts of results for monotone systems, we refer to [34, 56] and references therein.

Theorem 5.6.3 was established in [22] under the extra condition that m has at least one isolated global maximum. The new improvement is given in [26].

From the biological point of view, Theorem 5.6.3 is surprising at the first look. If $\mu < \nu$ and α is small, the species u always wins the competition. As α increases, the species u has the tendency to move toward more favorable regions, and so it has more competitive advantage than the species v and should still be the sole winner of the competition. However, the above theorem tells us that "smarter" species may not necessarily win the competition. A possible explanation for such coexistence is that as α becomes large, the "smarter" competitor moves toward and concentrates at places of locally more favorable environments, leaving enough room for the other species to survive in places with less resources.

It is shown in [22] that for any positive steady state solution (u_α, v_α) of (5.20–5.21), $\|u_\alpha\|_{L^2(\Omega)} \to 0$ as $\alpha \to \infty$, i.e., the total population size of species u becomes sufficiently small if α is large. It is natural to inquire whether the density function $u_\alpha \to 0$ in $L^\infty(\Omega)$ as $\alpha \to \infty$. As shown in our next result, the answer is negative in general. We will also show that u_α is concentrated at the global maximum of $m(x)$ in some cases.

Theorem 5.6.4 ([26]) *Suppose that $\int_\Omega m(x)\,dx > 0$ and all critical points of m are nondegenerate. Let $\mathcal{M}$ be the set of points of local maxima of m. Then for any positive steady state (u_α, v_α) of (5.20–5.21),*

$$\lim_{\alpha \to \infty} \max_{\overline{\Omega}} u_\alpha \geq \max_{\mathcal{M}}[m - \theta] > 0.$$

Assume further that $m(x)$ satisfies $\partial_n m \leq 0$ on $\partial\Omega$, and there exists $x_0 \in \Omega$ such that $\{x \in \overline{\Omega} : \nabla m(x) = 0\} = \{x_0\}$ and $\det(D^2 m(x_0)) \neq 0$. Then as $\alpha \to \infty$, $v_\alpha \to \theta$ in $C^{1+\beta}(\overline{\Omega})$ for every $\beta \in (0,1)$, and

$$\left\| u_\alpha\, e^{\alpha[\max_{\overline{\Omega}} m - m(x)]/\mu} - 2^{N/2}[m(x_0) - \theta(x_0)] \right\|_{L^\infty(\Omega)} \to 0. \qquad (5.23)$$

Theorem 5.6.4 shows that if m has a unique local maximum in $\overline{\Omega}$ which is also nondegenerate, then the "smarter" species is concentrated near this local maximum. For general m, we have the following

Conjecture. For sufficiently large α, (5.20–5.21) has a unique coexistence state, denoted by (u_α, v_α), which is globally asymptotically stable among nonnegative nontrivial initial data. Moreover, as $\alpha \to \infty$, u_α concentrates at all local maxima of $m(x)$ in $\overline{\Omega}$.

5.6.3 Evolutionarily Stable Dispersal Strategy

In this subsection we discuss the situation in which both competitors have a tendency to move upward along the resource gradient, and the equations that describe this competition model are

$$\begin{aligned}
u_t &= \nabla \cdot [\mu \nabla u - \alpha u \nabla m] + (m - u - v)u &&\text{in } \Omega \times (0, \infty), \\
v_t &= \nabla \cdot [\nu \nabla v - \beta v \nabla m] + (m - u - v)v &&\text{in } \Omega \times (0, \infty),
\end{aligned} \qquad (5.24)$$

with no-flux boundary conditions

$$[\mu \nabla u - \alpha u \nabla m] \cdot n = [\nu \nabla v - \beta v \nabla m] \cdot n = 0,$$

where α and β are nonnegative constants. The competitors represent different phenotypes of the same species which differ only in their dispersal mechanisms. We ask what type of dispersal strategy confers a competitive advantage. Our goal is to search for the regions of $(\alpha, \beta, \mu, \nu)$ in which either competition exclusion holds or stable coexistence occurs. A particularly interesting question is whether there exists one dispersal strategy which is evolutionarily stable, i.e., a population using it cannot be invaded by a small number of individuals of other competing species that uses a different strategy.

Even though the dispersal strategies for both species have same components, i.e., random movement and biased movement along resource gradients, they can be quite different. For instance, the species with stronger biased movement behaves like a specialist as it mainly pursues resources at places of locally most favorable environments. The species with moderate biased movement has a rather balanced dispersal strategy, and so it can be regarded as a generalist. It is fairly natural to ask the following:

Question. For arbitrary but fixed $\beta > 0$, do the two species always coexist if α is sufficiently large?

By previous coexistence results for the case $\beta = 0$, one might expect that the specialist and generalist should still coexist stably. However, we will show that the answer depends crucially upon the ratio β/ν, and at least two scenarios can occur: if the generalist's biased movement rate is relatively smaller than its own random movement rate, then indeed both species can coexist stably. However, if the biased movement of the generalist is relatively stronger than its random movement, then the generalist is always the winner, regardless of initial conditions. More precisely, we have the following two results that complement and contrast each other.

Theorem 5.6.5 ([24]) *Suppose that the set of critical points of $m(x)$ has measure zero. If $\beta/\nu \leq 1/\max_{\bar{\Omega}} m$, then for sufficiently large α, both semitrivial states are unstable, and (5.24) has at least one stable positive steady state.*

Theorem 5.6.5 is a generalization of the results in [22, 26] for the case $\beta = 0$. What happens if β/ν is suitably larger? It turns out that the answer is significantly different.

Theorem 5.6.6 ([24]) *Suppose that $m > 0$ in $\bar{\Omega}$, $\partial_n m < 0$ on $\partial\Omega$, m has only one critical point in $\bar{\Omega}$, denote by x_0, and x_0 satisfies $x_0 \in \Omega$ and $D^2 m(x_0) < 0$. If $\beta/\nu \geq 1/\min_{\bar{\Omega}} m$, then for sufficiently large α, the semi-trivial steady state $(0, \theta)$ is globally asymptotically stable.*

These results may have potential applications to the evolution of dispersal and population dynamics. In the context of dynamics of generalist and specialist, $\beta/\nu \leq 1/\max_{\bar{\Omega}} m$ implies that the two species can coexist stably for sufficiently large α, i.e., both generalist and specialist can coexist. However, for the case $\beta/\nu \geq 1/\min_{\bar{\Omega}} m$, competitive exclusion happens: the generalist always wins. That is, strong advection along resource gradients can be disadvantageous to the specialist and can even cause the extinction of the specialist in some circumstances. This seems to suggest that selection is for intermediate biased movement rate.

Two special but biologically relevant cases are worth mentioning:

(i) $\mu = \nu$ but $\alpha \neq \beta$. For this case, the competing species are identical in all aspects except their advection coefficients. When $\alpha > 0$ is small and $\beta = 0$, it is known that if Ω is convex then u is the winner in the competition. If α is large and $\beta = 0$, then both species can persist. For the general case $\alpha, \beta > 0$, we suspect that there is an evolutionarily stable dispersal strategy. Namely, there exists some α such that for any $\beta \neq \alpha$, the species v can not invade when rare.

(ii) $\mu \neq \nu$ but $\alpha = \beta$. For this case, the slower diffuser always wins if α is small. For suitably large values of α, very little is known.

Understanding these two special cases will shed light on more general situations, and can also be useful in studying other ecological problems, e.g., moving ranges of species and the effects of dispersal strategies on stream populations [91, 109].

It will also be interesting to consider N ($N \geq 3$) competing species and address similar questions. Suppose that all species are identical except their dispersal strategies.

(a) What happens if species 1 to $N - 1$ disperse only by random diffusion, and species N disperse by both random diffusion and directed movement towards habitat resources?

(b) What happens if all species have the same advection coefficients but different random dispersal rates?

(c) What happens if all species have same random dispersal rates but different advection coefficients?

We suspect that the answers to these questions are much more complicated for three or more species and rely on a solid understanding of the two-species case. Nevertheless, these are some of the interesting and challenging open problems on the evolution of conditional dispersal.

Acknowledgements

I thank Profs. Stephen Cantrell, Xinfu Chen, Chris Cosner, Vivian Hutson, Chiu-Yen Kao, Salome Martinez, Konstantin Mischaikow, Thomas Nagylaki, Wei-Ming Ni, Peter Poláčik, and Eiji Yanagida for the pleasant collaborations on the projects presented here and for sharing their amazing insights. I especially thank Steve Cantrell, Chris Cosner, and Vivian Hutson for reading this manuscript carefully and for their helpful comments. I thank Prof. Avner Friedman for inviting me to write this survey, and also thank him and Prof. Peter March for their warm encouragement. This work is partially supported by National Science Foundation grant DMS-0615845.

References

[1]. P.R. Armsworth and J.E. Roughgarden, The impact of directed versus random movement on population dynamics and biodiversity patterns, *Am. Nat.* **165** (2005) 449-465.

[2]. P.R. Armsworth and J.E. Roughgarden, Disturbance induces the contrasting evolution of reinforcement and dispersiveness in directed and random movers, *Evolution* **59** (2005) 2083-2096.

[3]. F. Belgacem, Elliptic Boundary Value Problems with Indefinite Weights: Variational Formulations of the Principal Eigenvalue and Applications, Pitman Research Notes in Mathematics, Vol. **368**, Longman, Harlow, U.K., 1997.

[4]. F. Belgacem and C. Cosner, The effects of dispersal along environmental gradients on the dynamics of populations in heterogeneous environment, *Canadian Appl. Math. Quarterly* **3** (1995) 379-397.

[5]. H. Berestycki, F. Hamel, and L. Roques, Analysis of the periodically fragmented environment model. I. Species persistence, *J. Math. Biol.* **51** (2005) 75-113.

[6]. D.E. Bowler and T.G. Benten, Causes and consequences of animal dispersal strategies: relating individual behavior to spatial dynamics, *Biol. Rev.* **80** (2005) 205-225.

[7]. K.J. Brown and S.S. Lin, On the existence of positive eigenvalue problem with indefinite weight function, *J. Math. Anal. Appl.* **75** (1980) 112-120.

[8]. R.S. Cantrell and C. Cosner, Diffusive logistic equations with indefinite weights: population models in a disrupted environments, *Proc. Roy. Soc. Edinburgh* **112A** (1989) 293-318.

[9]. R.S. Cantrell and C. Cosner, The effects of spatial heterogeneity in population dynamics, *J. Math. Biol.* **29** (1991) 315-338.

[10]. R.S. Cantrell and C. Cosner, Should a park be an island? *SIAM J. Appl. Math.* **53** (1993) 219-252.

[11]. R.S. Cantrell and C. Cosner, On the effects of spatial heterogeneity on the persistence of interacting species, *J. Math. Biol.* **37** (1998) 103-145.

[12]. R.S. Cantrell and C. Cosner, Spatial Ecology via Reaction-Diffusion Equations, Series in Mathematical and Computational Biology, John Wiley and Sons, Chichester, UK, 2003.

200   Y. Lou

[13]. R.S. Cantrell and C. Cosner, On the effects of nonlinear boundary conditions in diffusive logistic equations on bounded domains, *J. Diff. Eqs.* **231** (2006) 768-804.

[14]. R.S. Cantrell, C. Cosner, D.L. DeAngelis, and V. Padrón, The ideal free distribution as an evolutionarily stable strategy, *J. of Biological Dynamics*, to appear.

[15]. R.S. Cantrell, C. Cosner, and W.F. Fagan, Brucellosis, botflies and brinworms: the impact of edge habitats on pathogen transmission and species extinction, *J. Math. Biol.* **42** (2001) 95-119.

[16]. R.S. Cantrell, C. Cosner, and W.F. Fagan, Competitive reversals inside ecological preserves: the role of external habitat degradation, *J. Math Biol.* **37** (1998) 491-533.

[17]. R.S. Cantrell, C. Cosner, and W.F. Fagan, Habitat edges and predator-prey interactions: effects on critical patch size, *Math. Biosc.* **175** (2002) 31-55.

[18]. R.S. Cantrell, C. Cosner, and V. Hutson, Permanence in ecological systems with diffusion, *Proc. Roy. Soc. Edin.* **123A** (1993) 533-559.

[19]. R.S. Cantrell, C. Cosner, and V. Hutson, Ecological models, permanence and spatial heterogeneity, *Rocky Mount. J. Math.* **26** (1996) 1-35.

[20]. R.S. Cantrell, C. Cosner, and Y. Lou, Multiple reversals of competitive dominance in ecological reserves via external habitat degradation, *J. Dyn. Diff. Eqs.* **16** (2004) 973-1010.

[21]. R.S. Cantrell, C. Cosner, and Y. Lou, Movement towards better environments and the evolution of rapid diffusion, *Math Biosciences* **204** (2006) 199-214.

[22]. R.S. Cantrell, C. Cosner, and Y. Lou, Advection mediated coexistence of competing species, *Proc. Roy. Soc. Edinb.* **137A** (2007) 497-518.

[23]. A.N. Carvalho and J.K. Hale, Large diffusion with dispersion, *Nonl. Anal.* **17** (1991) 1139-1151.

[24]. X.F. Chen, R. Hambrock, and Y. Lou, Advection-induced coexistence and extinction in a two-species competition model, preprint, 2007.

[25]. X.Y. Chen, S. Jimbo, and Y. Morita, Stabilization of vortices in the Ginzburg-Landau equation with a variable diffusion coefficient, *SIAM J. Math. Anal.* **29** (1998) 903-912.

[26]. X.F. Chen and Y. Lou, Principal eigenvalue and eigenfunction of elliptic operator with large convection and its application to a competition model, *Indiana Univ. Math. J.*, accepted for publication, 2007.

[27]. Y.S. Choi, R. Lui, and Y. Yamada, Existence of global solutions for the Shigesada-Kawasaki-Teramoto model with weak cross-diffusion, *Disc. Cont. Dyn. Syst. A* **9** (2003) 1193-1200.

[28]. M. Conti, S. Terracini, and G. Verzini, A variational problem for the spatial segregation of reaction-diffusion systems, *Indiana Univ. Math. J.* **54** (2005) 779-815.

[29]. M. Conti, S. Terracini, and G. Verzini, Asymptotic estimates for the spatial segregation of competitive systems, *Adv. Math.* **195** (2005) 524-560.

[30]. E. Conway, D. Hoff, and J. Smoller, Large time behavior of solutions of systems of nonlinear reaction-diffusion equations, *SIAM J. Appl. Math.* **35** (1978) 1-16.

[31]. C. Cosner and Y. Lou, Does movement toward better environments always benefit a population? *J. Math. Anal. Appl.* **277** (2003) 489-503.

[32]. E.N. Dancer, Positivity of maps and applications. Topological nonlinear analysis, 303-340, Prog. Nonlinear Differential Equations Appl., **15**, edited by Matzeu and Vignoli, Birkhauser, Boston, 1995.

[33]. E.N. Dancer and Y. Du, Competing species equations with diffusion, large interactions, and jumping nonlinearities, *J. Diff. Eqs.* **114** (1994) 434-475.

[34]. E.N. Dancer, P. Hess, Stability of fixed points for order-preserving discrete-time dynamical systems, *J. Reine Angew. Math.* **419** (1991) 125-139.

[35]. J. Dockery, V. Hutson, K. Mischaikow, and M. Pernarowski, The evolution of slow dispersal rates: a reaction-diffusion model, *J. Math. Biol.* **37** (1998) 61-83.

[36]. M. Doebeli, Dispersal and dynamics. *Theor. Pop. Biol.* **47** (1995) 82-106.

[37]. C.P. Doncaster, J. Clobert, B. Doligez, L. Gustafsson, and E. Danchin, Balanced dispersal between spatially varying local populations: an alternative to the source-sink model, *Am. Nat.* **150** (1997) 425-445.

[38]. Y. Du, Effects of a degeneracy in the competition model, Part II. Perturbation and dynamical behavior, *J. Diff. Eqs.* **181** (2002) 133-164.

[39]. Y. Du, Realization of prescribed patterns in the competition model, *J. Diff. Eqs.* **193** (2003) 147-179.

[40]. Y. Du, Spatial patterns for population models in a heterogeneous environment, *Taiwanese J. Math.* **8** (2004) 155-182.

[41]. Y. Du, Bifurcation and related topics in elliptic problems. Stationary partial differential equations. Vol. II, 127–209, Handb. Differ. Equ., Elsevier/North-Holland, Amsterdam, 2005.

[42]. H. Evans, P. Kröger, and K. Kurata, On the placement of an obstacle or well to optimize the fundmental eigenvalue, *SIAM J. Math. Anal.* **33** (2001) 240-259.

[43]. W.F. Fagan, R.S. Cantrell, and C. Cosner, How habitat edges change species interactions: a synthesis of data and theory, *Am. Nat.* **153** (1999) 165-182.

[44]. W.H. Fleming, A selection-migration in population genetics, *J. Math. Biol.* **2** (1975) 219-223.

[45]. A. Friedman, Partial Differential Equations of Parabolic Type, Prentice-Hall, 1964.

[46]. J.E. Furter and J. López-Gómez, Diffusion-mediated permanence problem for a heterogeneous Lotka-Volterra competition model, *Proc. Roy. Soc. Edin.* **127A** (1997) 281-336.

[47]. D. Gilbarg and N. Trudinger, Elliptic Partial Differential Equation of Second Order, 2nd Ed., Springer-Verlag, Berlin, 1983.

[48]. S. Gourley and Y. Kuang, Two-species competition with high dispersal: the winning strategy, *Math. Biosci. Eng.* **2** (2005) 345-362.

[49]. J.K. Hale, Large diffusivity and asymptotic behavior in parabolic systems, *J. Math. Anal. Appl.* **118** (1986) 455-466.

[50]. J. K. Hale and G. Raugel, Reaction-diffusion equation on thin domains, *J. Math. Pures. Appl.* **71** (1992) 33-95.

[51]. J.K. Hale and K. Sakamoto, Shadow systems and attractors in reaction-diffusion equations, *Appl. Anal.* **32** (1989) 287-303.

[52]. I. Hanski, Metapopulation Ecology, Oxford Univ. Press, Oxford, 1999.

[53]. I. Hanski and C.D. Thomas, Metapopulation dynamics and conservation: a spatially explicit model applied to butterflies. *Biol. Conservation* **68** (1994) 167-180.

[54]. A. Hastings, Can spatial variation alone lead to selection for dispersal? *Theor. Pop. Biol.* **33** (1983) 311-314.

[55]. A. Hastings, Spatial heterogeneity and ecological models, *Ecology* **71** (1990) 426-428.

[56]. P. Hess, Periodic Parabolic Boundary Value Problems and Positivity, Longman Scientific & Technical, Harlow, UK, 1991.

[57]. M.W. Hirsch, Stability and convergence in strongly monotone dynamical systems, *J. Reine Angew. Math.* **383** (1988) 1-51.

[58]. M.W. Hirsch and H.L. Smith, Asymptotically stable equilibria for monotone semiflows, *Discrete Contin. Dyn. Syst. A* **14** (2006) 385-398.

[59]. E.E. Holmes, M.A. Lewis, J.E. Banks, and R.R. Veit, Partial differential equations in ecology: spatial interactions and population dynamics, *Ecology* **75** (1994) 17-29.

[60]. R.D. Holt, Population dynamics in two-patch environments: some anomalous consequences of an optimal habitat distribution, *Theor. Pop. Biol.* **28** (1985) 181-208.

[61]. R.D. Holt and M.A. McPeek, Chaotic population dynamics favors the evolution of dispersal, *Am. Nat.* **148** (1996) 709-718.

[62]. S. Hsu, H. Smith, and P. Waltman, Competitive exclusion and coexistence for competitive systems on ordered Banach spaces, *Trans. Amer. Math. Soc.* **348** (1996) 4083-4094.

[63]. V. Hutson, J. López-Gómez, K. Mischaikow, and G. Vickers, Limit behavior for a competing species problem with diffusion, in *Dynamical Systems and Applications*, World Sci. Ser. Appl. Anal. 4, World Scientific, River Edge, NJ, 1995, 501-533.

[64]. V. Hutson, Y. Lou, and K. Mischaikow, Spatial heterogeneity of resources versus Lotka-Volterra dynamics, *J. Diff. Eqs.* **185** (2002) 97-136.

[65]. V. Hutson, Y. Lou, and K. Mischaikow, Convergence in competition models with small diffusion coefficients, *J. Diff. Eqs.* **211** (2005) 135-161.

[66]. V. Hutson, Y. Lou, K. Mischaikow, and P. Poláčik, Competing species near the degenerate limit, *SIAM J. Math. Anal.* **35** (2003) 453-491.

[67]. V. Hutson, S. Martinez, K. Mischaikow, and G.T. Vickers, The evolution of dispersal, *J. Math. Biol.* **47** (2003) 483-517.

[68]. V. Hutson, K. Mischaikow, and P. Poláčik, The evolution of dispersal rates in a heterogeneous time-periodic environment, *J. Math. Biol.* **43** (2001) 501-533.

[69]. M. Iida, M. Mimura, and H. Ninomiya, Diffusion, cross-diffusion and competitive interaction, *J. Math. Biol.* **53** (2006) 617-641.

[70]. M. Iida, M. Tatsuya, H. Ninomiya, and E. Yanagida, Diffusion-induced extinction of a superior species in a competition system, *Japan J. Iudust. Appl. Math.* **15** (1998) 223-252.

[71]. J. Jiang, X. Liang, and X. Zhao, Saddle point behavior for monotone semiflows and reaction-diffusion models, *J. Diff. Eqs.* **203** (2004) 313-330.

[72]. Y. Kan-on and E. Yanagida, Existence of non-constant stable equilibria in competition-diffusion equations, *Hiroshima Math. J.* **23** (1993) 193-221.

[73]. C.Y Kao, Y. Lou, and E. Yanagida, Principal eigenvalue for an elliptic problem with indefinite weight on cylindrical domains, in preparation, 2007.

[74]. S. Kirkland, C.-K. Li, and S.J. Schreiber, On the evolution of dispersal in patchy environments, *SIAM J. Appl. Math.* **66** (2006) 1366-1382.

[75]. K. Kishimoto and H.F. Weinberger, The spatial homogeneity of stable equilibria of some reaction-diffusion systems on convex domains, *J. Diff. Eqs.* **58** (1985) 15-21.

[76]. K. Kurata and J. Shi, Optimal spatial harvesting strategy and symmetry-breaking, preprint, 2006.

[77]. K. Kurata, M. Shibata, and S. Sakamoto, Symmetry-breaking phenomena in an optimization problem for some nonlinear elliptic equation, *Appl. Math. Optim.* **50** (2004) 259-278.

[78]. C.L. Lehman and D. Tilman, Competition in spatial habitats. In: Tilman, D., Kareiva, P. (Eds.), Spatial Ecology. Princeton Univ. Press, Princeton, NJ, 1997, pp. 185-203.

[79]. S.A. Levin, H.C. Muller-Landau, R. Nathan, and J. Chave, The ecology and evolution of seed dispersal: a theoretical perspective, *Annu. Rev. Eco. Evol. Syst.* **34** (2003) 575-604.

[80]. J. López-Gómez, Coexistence and meta-coexistence for competing species, *Houston J. Math.* **29** (2003) 483-536.

[81]. J. López-Gómez and M. Molina-Meyer, Superlinear indefinite system beyond Lotka-Volterra models, *J. Diff. Eqs.* **221** (2006) 343-411.

[82]. Y. Lou, On the effects of migration and spatial heterogeneity on single and multiple species, *J. Diff. Eqs.* **223** (2006) 400-426.

[83]. Y. Lou, S. Martinez, and W.M. Ni, On 3 × 3 Lotka-Volterra competition systems with cross-diffusion, *Dis. Cont. Dyn. Syst. A* **6** (2000) 175-190.

[84]. Y. Lou, S. Martinez, and P. Poláčik, Loops and branches of coexistence states in a Lotka-Volterra competition model, *J. Diff. Eqs.* **230** (2006) 720-742.

[85]. Y. Lou and T. Nagylaki, Evolution of A Semilinear Parabolic System for Migration and Selection without dominance, *J. Diff. Eqs.* **225** (2006) 624-665.

[86]. Y. Lou, T. Nagylaki, and W.M. Ni, On diffusion-induced blowups in a co-operative model, *Nonl. Anal.: Theory, Meth. Appl.* **45** (2001) 329-342.

[87]. Y. Lou and W.M. Ni, Diffusion, self-diffusion and cross-diffusion, *J. Diff. Eqs.* **131** (1996) 79-131.

[88]. Y. Lou and W.M. Ni, Diffusion vs. cross-diffusion: an elliptic approach, *J. Diff. Eqs.* **154** (1999) 157-190.

[89]. Y. Lou, W.M. Ni, and S. Yotsutani, On a limiting system in the Lotka-Volterra competition with cross-diffusion. *Dis. Cont. Dyn. Syst. A* **10** (2004) 435-458.

[90]. Y. Lou and E. Yanagida, Minimization of the principal eigenvalue with indefinite weight and applications to population dynamics, *Japan J. Indus. Appl. Math* **23** (2006) 275-292.

[91]. F. Lutscher, E. Pachepsky, and M. Lewis, The effect of dispersal patterns on stream populations, *SIAM Rev.* **47** (2005) 749–772.

[92]. H. Matano, Existence of nontrivial unstable sets for equilibriums of strongly order-preserving systems, *J. Fac. Sci. Univ. Tokyo* **30** (1984) 645-673.

[93]. H. Matano and M. Mimura, Pattern formation in competition-diffusion systems in non-convex domains, *Publ. RIMS. Kyoto Univ.* **19** (1983) 1049-1079.

[94]. M.A. McPeek and R.D. Holt, The evolution of dispersal in spatially and temporally varying environments, *Am. Nat.* **140** (1992) 1010-1027.

[95]. M. Mimura, Stationary pattern of some density-dependent diffusion system with competitive dynamics, *Hiroshima Math. J.* **11** (1981) 621-635.

[96]. M. Mimura, S.I. Ei, and Q. Fang, Effect of domain-type on the coexistence problems in a competition-diffusion system, *J. Math. Biol.* **29** (1991) 219-237.

[97]. M. Mimura and K. Kawasaki, Spatial segregation in competitive interaction-diffusion equations, *J. Math. Biol.* **9** (1980) 49-64.

[98]. M. Mimura, Y. Nishiura, A. Tesei, and T. Tsujikawa, Coexistence problem for two competing species models with density-dependent diffusion, *Hiroshima Math. J.* **14** (1984) 425-449.

[99]. D.W. Morris, J.E. Diffendorfer, and P. Lundberg, Dispersal among habitats varying in fitness: reciprocating migration through ideal habitat selection, *Oioks* **107** (2004) 559-575.

[100]. N. Mizoguchi, N. Ninomiya, and E. Yanagida, On the blowup induced by diffusion in nonlinear systems, *J. Dyn. Diff. Eqs.* **10** (1998) 619-638.

[101]. J.D. Murray, Mathematical Biology II. Spatial models and Biomedical Applications, Interdisciplinary Applied Mathematics, Vol. **18**, 3rd ed. Springer-Verlag, New York, 2003.

[102]. C. Neuhauser, Mathematical challenges in spatial ecology, *Notices Amer. Math. Soc.* **48** (2001) 1304–1314.

[103]. W.M. Ni, Diffusion, cross-diffusion, and their spike-layer steady states, *Notices Amer. Math. Soc.* **45** (1998) 9-18.

[104]. W.M. Ni, Qualitative properties of solutions to elliptic problems. Stationary partial differential equations. Vol. I, 157-233, Handb. Differ. Equ., North-Holland, Amsterdam, 2004.

[105]. H. Ninomiya, Separatrices of competition-diffusion equations, *J. Math. Kyoto Univ.* **35** (1995) 539-567.

[106]. A. Okubo and S.A. Levin, Diffusion and Ecological Problems: Modern Perspectives, Interdisciplinary Applied Mathematics, Vol. **14**, 2nd ed. Springer, Berlin, 2001.

[107]. S. Pacala and J. Roughgarden, Spatial heterogeneity and interspecific competition, *Theor. Pop. Biol.* **21** (1982) 92-113.

[108]. P. Poláčik and E. Yanagida, Existence of stable subharmonic solutions for reaction-diffusion equations, *J. Diff. Eqs.* **169** (2001) 255-280.

[109]. A.B. Potapov and M.A. Lewis, Climate and competition: the effect of moving range boundaries on habitat invasibility, *Bull. Math. Biol.* **66** (2004) 975-1008.

[110]. M.H. Protter and H.F. Weinberger, Maximum Principles in Differential Equations, 2nd ed., Springer-Verlag, Berlin, 1984.

[111]. G. Raugel, Dynamics of partial differential equations on thin domains, *Dynamical systems* (Montecatini terme, 1994), 208-315, Lecture Notes in Math. **1609**, Springer, Berlin, 1995.

[112]. G. Raugel and G. Sell, Navier-Stokes equations on thin 3D domains. I. Global attractors and global regularity of solutions, *J. Amer. Math. Soc.* **6** (1993) 503-568.

[113]. J.C. Saut and B. Scheurer, Remarks on a nonlinear equation arising in population genetics, *Comm. Part. Diff. Eq.*, **23** (1978) 907-931.

[114]. S. Senn and P. Hess, On positive solutions of a linear elliptic boundary value problem with Neumann boundary conditions, *Math. Ann.* **258** (1982) 459-470.

[115]. N. Shigesada and K. Kawasaki, Biological Invasions: Theory and Practice, Oxford Series in Ecology and Evolution, Oxford University Press, Oxford, New York, Tokyo, 1997.

[116]. N. Shigesada, K. Kawasaki, and E. Teramoto, Spatial segregation of interacting species, *J. Theo. Biol.* **79** (1979) 83-99.

[117]. J.G. Skellam, Random dispersal in theoretical populations, *Biometrika* **38** (1951) 196-218.

[118]. H. Smith, Monotone Dynamical Systems. Mathematical Surveys and Monographs 41. American Mathematical Society, Providence, Rhode Island, U.S.A., 1995.

[119]. J.M.J. Travis and C. Dytham, Habitat persistence, habitat availability and the evolution of dispersal, *Proc. Roy. Soc. Lond. B* **266** (1999) 723-728.

[120]. J.M.J. Travis and D.R. French, Dispersal functions and spatial models: expanding our dispersal toolbox, *Ecology Letters* **3** (2000) 163-165.

[121]. P. Turchin, Qualitative Analysis of Movement, Sinauer Press, Sunderland, MA, 1998.

[122]. H.F. Weinberger, An example of blowup produced by equal diffusions, *J. Diff. Eqs.* **154** (1999) 225-237.

[123]. E. Yanagida, Existence of stable stationary solutions of scalar reaction-diffusion equations in thin tubular domains, *Appl. Anal.* **36** (1990) 171-188.

Lecture Notes in Mathematics

For information about earlier volumes
please contact your bookseller or Springer
LNM Online archive: springerlink.com

Recent Reprints and New Editions